KU-157-622

Online GIS and Spatial Metadata

David Green & Terry Bossomaier

London and New York

First published 2002 by Taylor & Francis
11 New Fetter Lane, London EC4P 4EE

Simultaneously published in the USA and Canada
by Taylor & Francis Inc,
29 West 35th Street, New York, NY 10001

Taylor & Francis is an imprint of the Taylor & Francis Group

© 2002 David Green & Terry Bossomaier

Printed and bound in Great Britain by
Biddles Ltd, Guildford and King's Lynn

All rights reserved. No part of this book may be reprinted or reproduced or
utilised in any form or by any electronic, mechanical, or other means, now
known or hereafter invented, including photocopying and recording, or in
any information storage or retrieval system, without permission in writing
from the publishers.

Every effort has been made to ensure that the advice and information in this
book is true and accurate at the time of going to press. However neither the
publisher nor the authors can accept any legal responsibility or liability for
any errors or omissions that may be made. In the case of drug
administration, any medical procedure or the use of technical equipment
mentioned within this book, you are strongly advised to consult the
manufacturer's guidelines.

Publisher's Note
This book has been prepared from camera-ready copy provided by the
authors.

British Library Cataloguing in Publication Data
A catalogue record for this book is available from the British Library

Library of Congress Cataloging in Publication Data
Bossomaier, Terry R.J. (Terry Richard John)
    Online GIS and spatial metadata/ Terry Bossomaier & David Green.
    p. cm.
  Includes bibliographical references (p.).
1.Geographic information systems. 2.World Wide Web. 3. Metadata.
I.Title: Online geographic information systems and spatial metadata.
II. Green, David, 1954 Aug. 9- III. Title.

G70.212 .B69 2001   2001042290
910'.285--dc21

ISBN 0-748-40954-8

Learning Resources
Centre
1222298 4

WA 1222298 4

Online GIS Metadata

7·12·01

Contents

Preface

This book arises out of two areas of interest that we have shared over many years. One is the environmental informatics, and the many questions, technological and practical, involved in collating, interpreting and disseminating environmental information, which is essentially geographic in nature. The other is our involvement with Web technology, with its potential for providing a standard interface for many kinds of information, and for seamlessly integrating information resources on a scale never seen before.

A major part of the above research has included experimentation with methodologies for online GIS and for automating distributed information systems. Many of the examples presented herein are derived from that work.

Despite the evident potential for online GIS, for many years it remained a curiosity rather than a core technology. The stimulus for writing this book now was a conference that we organised in 2000 on the topic. During the course of that meeting it became clear that interest in this field was finally beginning to pick up. The time seemed ripe to provide an overview of the field.

In this book we have tried to achieve several goals. One is to provide an overview of the field that goes beyond the superficial. We aim to provide a starting point for professionals who are trying to learn about the technology, or who need to set up and run their own online services. We expect that readers will include both GIS professionals who wish to learn more about the issues raised by the Web, and Web professionals who wish to learn about what's involved in running GIS in an online environment.

This is not a text about GIS *per se*. There are many other excellent books that already cover the basic theory and practice of GIS more than adequately. However, for programmers and others new to geographic information, we do provide a brief primer on the basics.

Conversely, for GIS people new to Web technology, we have likewise tried to provide an introduction to the working of HTTP and other matters. Although we delve into deeply technical matters, we have nevertheless tried to make the account readable. The technical details are mainly to show readers examples of real scripts, markup and other elements that make the technology work. We have tried to avoid turning the book into a technical manual. We would hope that our accounts of the various topics is clear and lucid enough to help non-technical readers understand all the issues. We encourage anyone who finds source code daunting to simply skip over the scripts.

Another goal of this book is to try to sketch out some ideas for future directions. The entire last chapter is devoted to describing several foreseeable developments. However, we have also dealt with several issues much earlier. One example is our object-oriented approach, which we sketch out, immediately in Chapter 1. Most GIS systems and standards are currently not object-oriented, although the approach seems completely natural and ideally suited for GIS, especially in an online, distributed environment.

An indication of the growing activity in online GIS and related technology is the rapid rate at which protocols, standards and software are changing. One of the greatest difficulties we faced was that the ground kept shifting even while we were writing! We had to go back and revise several chapters because they had already gone out of date since we wrote them.

Given the alarming speed at which the field is changing, we have been careful to ensure that the finished text would remain current for a reasonable time. To achieve this goal we have taken several steps. The first is to be less prescriptive about methods and details. We have deliberately avoided describing commercial products, except as examples to indicate directions that the technology is taking. For one thing, this information will be readily available. For another, products are still changing rapidly.

Our experience as educators in computing and information technology is that people can use technology better if they understand the basics. So in Chapter 3, for instance, we have aimed to help readers understand how online data processing works and just what is involved in getting simple geographic services to run.

The second step we have taken is to establish a Web site where readers can access further information. This site

```
http://clio.mit.csu.edu.au/smdogis/
```

will provide a range of services to supplement the book. These facilities include:

- working examples of the scripts etc;
- links to current versions of the resources we describe;
- links to other sites and services concerning online GIS;
- link to an online journal where researchers can publish recent work.

In conclusion, we hope that readers will find this book useful. For those of you new to this field, we hope you find it as exciting and rewarding as we do.

David G. Green and Terry R. J. Bossomaier

Albury Australia, January 2001

Acknowledgements

Finally, we are indebted to the following organizations for granting permission to reproduce sample pages from their online services:

- Museum of Victoria (Fig. 1.1)
- Pierce County, Washington USA (Fig. 2.1) MAP-Your-Way™
- Xerox PARC Map Viewer (Fig. 2.2 is courtesy of Xerox, Palo Alto Research Center)
- US Census Bureau (Fig. 2.4, 2.5) TIGER mapping system
- Environment Australia (Fig. 4.7) Australian Atlas
- World Data Centre (Fig. 4.11) Webmapper
- Mapquest (Fig. 11.1)
- The Metadata examples of ANZLIC standards provided in Section 9.1 is Copyright © Commonwealth of Australia, AUSLIG, Australia's national mapping agency. All rights reserved. Reproduced by permission of the General Manager, Australian Surveying and Land Information Group, Department of Industry, Science and Resources, Canberra, ACT. Apart from any use as permitted under the Copyright Act 1968, no part may be reproduced by any process without prior written permission from AUSLIG. Requests and enquiries concerning reproduction and rights should be addressed to the Manager, Australian Surveying and Land Information Group, Department of Industry, Science and Resources, PO Box 2, Belconnen, ACT, 2616, or by email to copyright@auslig.gov.au

This book would not have been possible without the support of the many individuals and organisations who helped us. In Chapter 9 we have made extensive use of material provided to us by Dr Hugh Caulkins. In Chapter 10 we have made use of material, including several figures, kindly provided by David Newth. We are also grateful to him for development of the SLEEP scripting language, which we use to provide examples of script processing in the latter part of Chapter 3. In that chapter also, we are grateful to Tony Steinke, who wrote the original version of Charles Sturt University's Map Maker service, and to Paul Bristow, who has developed and maintained the service for several years now.

We are also grateful to colleagues for their assistance with examples and other material in the book. Some other examples described were implemented and maintained by Larry Benton. Some of the javascript examples presented in Chapter 4 were based on material that Larry developed for the Guide to Australia. In Chapter 4, too, we thank Darren Stuart, who developed the java demonstrations presented. We are grateful to Joanne Lawrence for her assistance with the production of the book and to Simon McDonald for providing valuable comments on a draft of the manuscript.

Finally, DGG is grateful to Charles Sturt University for funds that helped with the development of some of the material, and for the opportunity to write portions of the text during study leave.

CHAPTER 1

Perspectives on global data

1.1 GEOGRAPHIC INFORMATION SYSTEMS

Geographic information systems (GIS) are computer programs for acquiring, storing, interpreting, and displaying spatially organised information. GIS had its origins in many different disciplines, including electronic cartography, geological surveys, environmental management, and urban planning. It has now become an essential tool in all of these professions, as well as many others.

At the time this book is being written a great transition is taking place in the way geographic information systems function. Instead of being isolated in stand-alone machines, a new environment is gradually being created in which geographic information is stored and accessed over the Internet. In this book we argue that this transition has implications that go far beyond a different format. One of the effects is the greater accessibility that the Internet offers. However, even more significant is the potential to combine information from many different sources in ways that were never previously possible. To work in the new environment, managers, developers and users all need to learn the basic technology involved. This book is a response to that need.

Many commercial developments in online mapping are now in progress. Our goal is not to describe or catalogue them in any detail, but to examine the issues underlying their development. However, before we can consider the issues involved in placing geographic information systems online, it is essential to have a clear picture of the kinds of information and functionality involved. The following section briefly reviews some of the main concepts in GIS.

1.2 A BRIEF PRIMER ON THE NATURE OF GIS

1.2.1 GIS data

Geographic data consists of layers. A *layer* is a set of geographically indexed data with a common theme or type. Examples might include coastlines, roads, topography, towns, public lands. To form a map, the GIS user selects a *base map* (usually a set of crucial layers, such as coastlines, or roads) and overlays selected layers. Layers are of three types:

❑ *vector layers* consist of objects that are points (e.g. towns), lines (e.g. roads) or polygons (e.g. national or state boundaries). Data of this kind are usually stored in database tables. Each record in the table contains attributes about individual objects in space, including their location.

❑ *raster layers* consist of data about sites within a region. Examples include satellite images, digitised aerial photographs and quadrat samples. The region is divided into a grid of cells (or *pixels*), each representing an area of the land surface. The layer contains attributes for each cell. For instance in a satellite image the attributes might be a set of intensity measurements at different light frequencies or a classification of the land features within the cell (e.g. forest, farmland, water).

❑ *digital models* are functions that compute values for land attributes at any location. For instance a *digital elevation model* would interpolate a value for the elevation, based on values obtained by surveying. In practice, digital models are usually converted to vector or raster layers when they are to be displayed or otherwise used.

As mentioned above, the data in vector layers are often stored in database tables. This enables a wide range of searching and indexing by attributes. The data may also be indexed spatially. For instance a *quadtree* is a construct that divides a region successively into four quadrants, down to any desired scale. Any item is indexed by the quadrants in which they li.e. This procedure (the spatial equivalent of a binary search tree) allows for rapid searching by location.

Overlays form a classic analytic tool in GIS. Some online services allow data layers to be overlaid on various backgrounds. For instance, the Victorian Butterfly Database (Fig. 1.1) allows the user to plot the point distributions of species against a variety of environmental classifications, each of which provide clues about factors controlling the species distributions.

1.2.2 GIS functionality

Although most commonly associated with map production, GIS encompasses a wide range of forms and functions. The most common (but far from only!) outputs from a GIS are maps. These maps are either the result of a simple plot of layers for a given region, or they display the results of a query. Queries involving GIS can be classified as first or second order in nature.

First order questions involve features within a single layer. Examples include the following:

❑ Where is object X? e.g. Where is Sydney?
❑ How big (or long) is object X? What is the area of Europe?
❑ Where are all objects having a certain property? Where are oil wells in Alberta?

Second order questions involve overlays of two or more layers, such as:

❑ What objects are found within areas of a given type? e.g. What mines lie within a state or district?
❑ What is the area of overlap between two kinds of objects? For example, what areas of rainforests lie within national parks? What farms does a proposed highway run through?
❑ What environmental factors influence distribution of a rare species?

The process of *overlay* often involves the construction of a new data layer from existing ones. Suppose that two layers consist of polygons (e.g. electoral districts and irrigation regions). Then the overlay would consists of polygons formed by the intersection of polygons in the two layers. For instance in the example given, these polygons would show the areas within each electorate that were subject to irrigation.

Two important kinds of functions within a GIS are data analysis and models. Data analysis tools include a host of techniques for comparing and analysing data layers. Some examples include:

- *kriging* (e.g. estimating size of ore bodies) and other methods of interpolation (e.g. elevation);

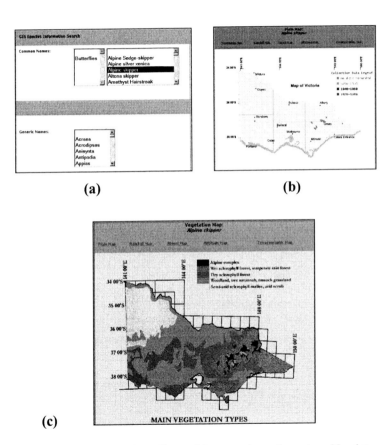

Figure 1.1. An example of an online spatial query and mapping system. Mapping of species distributions provided by the Museum of Victoria's informatics service. (a) A menu is used to select species, (b) A resulting plot of specimen locations, and (c) An overlay of sites on a map of vegetation types. Source: http://www.mov.vic. gov.au/bioinformatics/butter/#Species Mapper .

❑   spatial correlation of variables (e.g. incidence of medical conditions versus pollution levels by district);
❑   nearest neighbour analysis (e.g. the numbers of seabird species that nest close together);
❑   fractal dimension of shapes (e.g. forest boundaries).

Models in GIS are usually layers obtained by manipulation of other layers. For instance, operations on maps of topography, streams and rainfall, and soil type might be inputs to a model whose output is a map of soil moisture. Traditionally these layers have all been handled by a single integrated package. Yet at the same time, it is clear that these datasets might come from different places. The online paradigm would allow each of these components to be dynamically accessed from different sites.

1.2.3 The object model of GIS

One of the impacts of GIS was to introduce a whole new way of looking at spatial data. Although cartography had been operating digitally for years before GIS appeared on the scene, the emphasis was solely on producing maps. A road map, for instance, would consists of lines to be drawn on the page. Names for roads and towns formed a separate set of textual data. There was no direct link between (say) the point representing a town and the name of that town. These associations became apparent only when the map was printed.

As scientists started to use spatial data for study purposes, the drawbacks of the map-oriented approach rapidly became obvious, as the following examples illustrate:

❑   The data for the rivers in a region were stored as dozens of separate line segments. The data for individual rivers consisted of many isolated line segments, with no indication that the segments formed a single feature. The rivers were broken into segments because the map required gaps wherever other features (e.g. towns, roads) were to be drawn over the top.
❑   In a gazetteer of towns, each location was found to have a small error. These errors turned out to be offsets that were added to allow each place name to be plotted in a convenient space on the map, not at the actual location.

The biggest change that GIS introduces is to store geographical features as distinct **objects**. So a road would be stored as an object with certain attributes. These attributes would include its name, its type (i.e. a road), its quality (e.g. a highway) and a sequence of coordinates that represent the path that the road follows across the landscape. For instance, the map shown (Fig. 1.2) consists of the following objects (Table 1.1).

Table 1.1. Data for the objects shown in Fig. 1.2.

Object	Name	Type	Attribute
1	Sydney	Town	Capital city
2	Bathurst	Town	City
3	Blue Mts H'way	Road	Highway

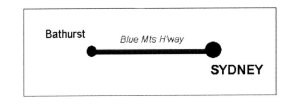

(a)

(b)

Figure 1.2. Simple geographic objects: a road joining two cities. The cities have two attributes (name and size), as also does the road (name and grade). (a) The objects plotted as they might appear on a map. (b) A simple model of the two classes of objects, drawn using the Universal Modelling Language (UML). The boxes denote a class of object. Associated with each class of object, the model defines two attributes (e.g. name, size for towns) and a single method (e.g. plot_as_a_circle). The line joining the classes denotes a relationship between them. The numbers at either end of the line (1..*) indicate that each road is related to one or more towns, and vice versa.

In this **object-oriented** approach, each geographic object belongs to some **class** of objects. In Figure 1.2, the objects Sydney and Bathurst would both belong to the class *city*; and each of them is termed an *instance* of the city class (Fig. 1.3). The notion of classes is a very natural one in GIS. It also has implications for the way geographic data is treated.

Each object has attributes associated with it. For the city class, these would at least include a name and a location (latitude and longitude). However, it might also include many other *attributes* that we could record about a city, such as its population, the state in which it lies, the address and phone number of the city offices, and so on. The importance of defining a class is that for any other city that we might want to add to the GIS, we have a precise list of data that we need to supply.

The object approach to GIS has many advantages. The first is that it encapsulates everything we need about each geographic entity. This not only includes data, but also extends to **methods** of handling that data. So, for instance, the city object might include a method for drawing a city on a map. In this case it would plot a city as a circle on a map, with the size of the circle being determined by the population size of city.

Another advantage of objects, is that we can define relationships between different classes of objects. For instance, the class of cities that we described above really belongs to a more general class of objects that we might term "inhabited places". This general class would not only include cities but also towns, military bases, research stations, mining camps, farms and any other kinds of places that humans inhabit.

The city class inherits some of its attributes from the general class of inhabited places. These basic properties would include the location, as well as (say) the current population. Other properties of the city class, such as contact details for the city offices, would be unique to it, and would not be shared by all inhabited places. The inhabited places actually form a hierarchy (Fig. 1.3), with capital cities *inheriting* properties from cities, cities inheriting attributes and methods from towns, and towns inheriting attributes from inhabited places Table 1.2. Thus a capital city has six attributes, all but one of which it inherits from the more general classes to which it belongs.

Object models are usually described in terms of classes of objects and the relationships between them (Fig. 1.3). The Uniform Modelling Language (UML) provides a standard approach for specifying object models of data and processes (Larman, 1998).

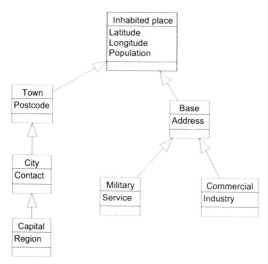

Figure 1.3. A simple data model, showing the relationships between several classes of geographic objects, drawn using the Universal Modelling Language (UML). Attributes are shown for each object, but methods are omitted. The arrows indicate general-special (Genspec) relationships between classes of objects. A city, for instance is a special kind of town and inherits all the attributes of a town. Likewise a capital inherits attributes of both (i.e. postcode, contact and region).

Table 1.2. The class hierarchy for inhabited places.

Class	Attributes	Methods
Inhabited place	Latitude Longitude Population	Draw a small circle
Town	Postcode	Draw a circle
City	Contact for city offices	Draw a shaded circle
Capital	Region served	Draw a large shaded circle

Another important type of link between classes of objects is a *whole-part relationship*. A map, for instance, consists of several elements. There are the border, the scale and possibly several layers, each containing many elements. Figure 1.4 shows a simple example of a class diagram showing the whole map as an aggregate of parts. So, for instance, to plot a whole map, you need to plot each part in turn.

A further interesting area of activity is the object-oriented database. Traditional relational databases store data as tables and relationships between them. Until the surge of interest in multimedia and the web over the last decade, they had been more or less limited to textual information. In attempting to store other material such as video or images, considerable changes were required in storage and query mechanisms.

Around the same time, the idea of storing data as complete objects with an associated object-query language arose. In the relational model a complex object might be broken up and split over many tables. In the object model the object (data and methods) are stored as an entity. An example might be a map and associated methods for extracting distances between places and many other things.

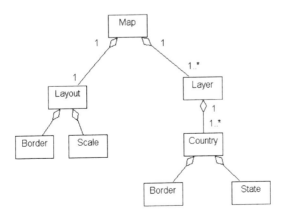

Figure 1.4. A map class is an aggregate consisting of several parts.

The OO database model is certainly very attractive for GIS style data. However, the commercial reality has been different. Enterprise databases store vast amounts of information, require exceptional robustness and stability and embody much experience in tuning and optimisation. OO starts a long way behind, and the relational databases have taken on board some of their advantages such as stored procedures, while hybrid object-relational databases have evolved. Thus OO databases are still a relatively small part of the market.

1.3 THE RISE OF THE INTERNET

The Internet (Krol, 1992) is a vast communications network that links together more than 2 million computers all over the world.

The rapid growth of Internet activity over the last few years has produced a literal explosion of information. From the user's point of view this process has emphasised several crucial needs:

- ❑ *Organization* Ensuring that users can obtain information easily and quickly. Various projects have developed general indexes of pointers to network information services (especially Internet search engines). However, these struggle to cope with the sheer volume of available information. Self-organization, based on user interests and priorities, is the only practical solution.
- ❑ *Stability* Ensuring that sources remain available and that links do not go "stale". Rather than gathering information at a single centre, an important principle is that the site that maintains a piece of information should be the principal source. Copies of (say) a dataset can become out-of-date very quickly, so it is more efficient for other sites to make links to the site that maintains a dataset, rather than take copies of it.
- ❑ *Quality* Ensuring that information is valid, that data are up-to-date and accurate, and that software works correctly.
- ❑ *Standardisation* Ensuring that the form and content of information make it easy to use.

The rise of the Internet, and especially the World Wide Web during the 1990s, revolutionised the dissemination of information. Once an organisation had published information online, anyone, anywhere could access it. This capacity meant that people could access relevant information more simply than in the past. They could also access it faster and in greater quantities. It also raised the potential for organisations to provide wider access to information resources that normally require specialised software or hardware. For example by filling in an online form, remote users can query databases. Geographic information systems that previously required specialised, and often expensive, equipment, can now be accessed remotely via a standard Web browser.

The volume of information available on the World Wide Web has grown exponentially since 1992, when the National Centre for Supercomputer Applications (NCSA) first released a multimedia browser (Mosaic). This explosion was driven first by data providers who recognised the potential audience that published information could attract. As the volume of information grew, users began to drive the process by demanding that information be available online.

The explosion of online information is a problem: finding one item amongst millions is akin to finding a needle in a haystack. Potential solutions for searching include the use of intelligent agents that continually sift and record relevant items and the promotion of metadata standards to make documents self-indexing. Information networks (see below) provide a way of organising sources of information.

The Internet is governed by the Internet Society (ISOC), which includes technical committees to vet and oversee the development and implementation of new standards and protocols. The World Wide Web is now governed by the World Wide Web Consortium (W3C), and has been actively developing many of the standards that we touch on in this book.

1.4 ADVANTAGES OF DISTRIBUTED INFORMATION SYSTEMS

Perhaps the greatest impact of the Internet is the ability to merge information from many different sources in seamless fashion (Green 1994). This ability opens the prospect of data sharing and cooperation on scales that were formerly impossible. It also brings the need for coordination sharply into focus.

As a geographic information system, the World Wide Web has some important advantages. One of the greatest practical problems in the development of geographic information systems is the sheer volume of data that needs to be gathered. Simply gathering datasets from suppliers can be a long drawn out process. Most systems require specialised data that the developers have to gather themselves. Inevitably, the lack of communication between developers leads to much duplication of effort. The Internet has the potential to eliminate these problems.

In principle, the suppliers of individual datasets or data layers could distribute their products online. This approach would not only speed up development of any GIS, but it would also help to eliminate duplication. In effect, online distribution of data would have the effect of increasing the potential volume of information available to GIS developers. It would distribute the workload amongst many organisations. This would also simplify the updating of information.

Another advantage of online GIS is that it expands the potential pool of GIS developers and users. As we shall see in Chapter 2, there are many options for placing GIS online. The result is that the overall costs can be much less for developers. It is not only possible, but also cost effective to implement small GIS that are designed for particular limited applications, such as providing a geographic interface to online documents or databases.

For some purposes, it is not even necessary for online publishers to implement a GIS at all. Instead, GIS services can be provided by leasing information from a specialist GIS site. For instance, suppose that a travel agency that provides information about (say) tourist attractions, accommodation and so on, wants to provide street maps to show the location of each site in its database. This is a very useful service for travellers who are trying to find their hotel in a strange city. Rather than developing a series of maps itself, the agency could arrange to link to sites that provide the necessary street maps. Such arrangements lead to many new commercial prospects for GIS, chiefly through on-selling of geographic services. We return to some of these new commercial models in Chapter 11.

For GIS users the prospects are equally exciting. The Internet brings GIS within reach of millions of users who previously could not afford the necessary equipment and specialised software. Apart from the availability of numerous free services, there is also the potential for access to a fully fledged GIS on demand. For instance, instead of buying an entire GIS themselves, users could buy GIS services from providers as they require them. For regular users, these services could take the form of subscription accounts to an online commercial system. On the other hand, irregular, one-time users could buy particular services in much the same way as they might previously have bought a paper map.

To realise the above possibilities, managers of online GIS will need to develop a climate of cooperation and sharing that is still foreign to many. The Internet is an ideal medium for collaboration. It makes possible communication and information sharing on scales that were hitherto unheard of. However, the explosive growth of the Internet at first led to enormous confusion. Organisations duplicated services and facilities in inconsistent ways. This pattern was exacerbated by commercial interests, which saw the Internet as an extension of their traditional competitive marketplace. The essential obstacle to information sharing is the tension between self-interest and cooperation. To resolve these issues, organisations and nations need to agree on protocols and standards for data recording, quality assurance, custodianship, copyright, legal liability and for indexing. In later chapters (especially Chapter 6) we will examine these issues in detail.

1.5 EXAMPLES OF DISTRIBUTED INFORMATION

The ability of the World Wide Web to link information from many different sources creates a synergy effect in which the overall information resource may be greater than the sum of its parts. One of the earliest demonstrations of this power was a service called The Virtual Tourist, created by Brandon Plewe (1997). It consisted of a central geographic index (point and click on a map) to access information about different countries. However, the important thing was that the information was distributed across literally thousands of sites. The geographic index merely pointed to country indexes, which in turn pointed to sources of detailed information about particular topics. The result was a system that allowed users to zoom in to any country and obtain detailed information about tourism, weather, and other useful information.

There are many examples, in many different fields of activity, that highlight the advantages of distributed, online information systems. Many of these have been in fields that relied on the Internet from the beginning. One such field is biotechnology. The vast online resources are compilations of genomic and protein sequences, enzymes, type cultures and taxonomic records. There are also many large repositories of useful software and online services for processing and interpreting data. Most of the prominent sites provide a wide range of other information as well, including bibliographies, electronic newsgroups and educational material. All of the resources are accessible online. Some of the larger facilities, such as the European Molecular Laboratory, are actually networks of collaborating organisations. The field has even reached the stage where many funding agencies and scientific journals actually require researchers to contribute their data to online repositories as a condition of acceptance of a grant or paper.

Many kinds of environmental information are already online. As scientists, we have been involved ourselves in some of the efforts already in progress to compile online information about many kinds of environmental resources, such as forestry and global biodiversity. In each case, the new possibilities are leading people and organisations to rethink the way they do things, to look at the broader geographic context, and to initiate schemes to enhance international cooperation.

1.6 OVERVIEW OF THIS BOOK

In keeping with our view that online GIS represents a new working environment, we attempt to achieve two goals in this book. The first is to give readers an overview of the basic technology involved in online geographic information systems. The second is to outline models for how the development of online geographic information might be coordinated.

There are several different aspects we need to consider. First, we have to understand the mechanisms of GIS processing over the Web. We then have to move on to consider how data is organised, accessed, searched, maintained, purchased and processed online. This involves us tackling some fairly complex standards, which are currently redefining the way in which online GIS will operate in the future. The first of these is XML, which underlies just about all of the current web standards. One such crucial standard is RDF the metadata standard for the web. The second major concept we need to tackle is that of distributed objects, and how it fits into the OpenGIS framework.

With all of the above details in place, we can now look at the content of the metadata standards for GIS information and proceed to look at how we do things online in a serious way.

One area we do avoid is detailed discussion of existing packages for Web GIS. The chief reason for this omission is that the nature and range of tools is likely to change rapidly in the near future as vendors respond to developments with the relevant standards.

Chapters 2 to 5 introduce the technical methods involved in developing and implementing geographic information systems on the Internet. They are intended to provide potential developers with technical details and examples to help them understand the issues. They also provide the technical background that readers need to appreciate many of the issues discussed in later chapters. Chapter 2 outlines the main technical issues surrounding online GIS and gives an overview of the options available. Chapter 3 describes the kinds of facilities and operations that can be implemented on a Web server. Chapter 4 describes GIS tools that can be implemented to operate on a standard Web browser.

In Chapter 5 we get to a key development, XML. For many years, the SGML international standard languished, little used outside a few big organisations such as the US military. Partly this was due to its complexity and the high cost of processing tools. Partly the superficial attractiveness of WYSIWYG publishing pushed it into the shadows. But as the web grew, the need for better organisational and searching methods came to the fore. HTML, the language of the web, is in fact an SGML DTD, and it became apparent that a simpler version of the full SGML standard would be advantageous. So, XML came about and has vacuumed up most of the other web standards. SGML also became the choice for writing spatial metadata standards, but is now being replaced by XML.

Chapters 6 to 11 examine technical issues involved in coordinating the development of geographic information in the Internet's distributed environment. These issues include many technical matters, but they also concern the ways in which human collaboration impinge on technology. Chapter 6 begins the discussion by outlining the nature of information networks, that is systems in which information is distributed across many different sites. Chapter 7 outlines some of the interoperable standards involved in distributed information systems. The biggest such initiative in the spatial arena is the OpenGIS consortium, which is essentially designing vendor independent standards for many spatial operations. The particular standards we look at in Chapter 7 relate to distributed objects, in which the location of objects on a computer network is essentially transparent.

Chapter 8 looks at the conceptual framework of metadata, by studying the RDF and similar standards for the Web. Chapter 9 follows this by describing several metadata standards in use around the world for spatial metadata. This chapter builds on the XML work in Chapter 5, the distributed object model of Chapter 7 and the metadata fundamentals in Chapter 8. Chapter 10 looks at ways in which distributed information can be built into data warehouses, and introduces basic ideas in data mining within such systems.

The final part of the book, Chapters 10 and 11, look at the prospects for the future development of online GIS. Chapter 10 discusses some of the emerging new technologies associated with geographic information. Chapter 11 looks at the possibilities inherent in a global geographic information system, as well as some of the possibilities raised by online GIS.

CHAPTER 2

GIS and the Internet

2.1 THE ADVANTAGES OF ONLINE GIS

The World Wide Web is fast becoming a standard platform for geographic information systems (GIS). A vast range of geographic information services already exists on the World Wide Web (Green, 1998) and the range of environmental information now available online is impressive. Commercial developers are already producing on-line versions of their GIS software. Many on-line services include spatial queries or map drawing.

Online GIS has several potential advantages over stand-alone systems. These advantages include:

❑ *World-wide access* An information system on the Web is accessible from anywhere in the world.
❑ *Standard interface* Every Web user has a browser, so any system that uses the Web is accessible by everyone, without the need for costly, specialised equipment.
❑ *Faster, more cost-effective maintenance* Information can be accessed at its source, so there is less need to collate data at a central location.

In this chapter we set out to do the following:

1. Examine some of the technical issues involved in developing an online GIS, and some of the options for dealing with them.
2. Briefly look at some examples of existing geographic information systems that are online.
3. Demonstrate some simple methods for implementing basic types of systems.
4. Examine some issues for the future development of online GIS.

2.2 ISSUES ARISING IN THE NEW MEDIUM

The Internet, and in particular the World Wide Web, is a new environment for all kinds of computing. As such it raises many issues and poses many challenges that do not exist in a stand-alone environment.

In the following discussion of online GIS, we shall for the most part gloss over methods of implementing standard GIS operations. Instead we focus on issues involved in placing the GIS online. The main issues that have to be addressed arise from:

1. the nature of the Web environment, and
2. the separation of the user interface from geographic data and processing.

2.2.1 The Web environment

The Web uses the Hypertext Transfer Protocol (HTTP) to communicate queries and responses across the Internet. HTTP is a *client-server protocol*. This means that a "client" – the user's browser program – transmits a query to a Web server, which then sends back its response. In HTTP version 1.0 (which is still used by most servers at the time of writing), these transactions are carried out on a connectionless, single query basis. Even if a user makes a series of queries to the same server, the server normally retains no history of past queries, and no client "session" is established. This is in contrast to several other Internet protocols, such as FTP and Telnet, where the server establishes a dialogue with clients who "log in".

Interactive GIS software that runs on a dedicated machine makes the implicit assumption that the program's current state is a direct result of a user's interactions with it. Remote GIS cannot do this under HTTP1.0, which necessitated work-arounds such as using hidden fields and caching.

HTTP and associated client/server software have the advantage of being a generic technology. Any service provided via the Hypertext Transfer Protocol is immediately available to anyone, on any type of machine, who is running a suitable client program.

Important features of WWW browsers and clients include:

❑ they permit browsing of ALL of the main network protocols (FTP, Telnet, etc.);

❑ they permit both text formatting and images that are embedded directly within text, thus providing the capability of a true "electronic book";

❑ they integrate freely available third party display tools for image data, sound, Postscript, animation, etc.;

❑ they permit seamless integration of a user's own local data (without the need of a server) with information from servers anywhere on the Web;

❑ the *forms interface* allows users to interact with documents that appear as forms (including buttons, menus, dialog boxes) which can pass complex queries back to the server;

❑ the *imagemap interface* allows users to query a map interactively. This would allow (say) a user to get information about different countries by clicking on a world map, in GIS-like fashion;

❑ the *authorisation* feature provides various security options, such as restricting access to particular information, passwords etc.;

❑ the *SQL gateway* allows servers to pass queries to databases. Such gateways are already implemented for many databases (e.g. flora and fauna of Europe and the Americas, DNA sequences);

❑ the ability to run scripts or programs on the server and to deliver the results to WWW;

❑ the ability to include files dynamically and thus build up and deliver documents "on the fly".

These generic technologies provide enough functionality to do simple spatial queries. Two more recent web innovations promise much more:

❑ *Java* from Sun Microsystems is a fully fledged programming language (see Chapter 4) with advanced graphics features, soon to be fully three-dimensional. It is machine independent and (at least with experience so far) secure.

❑ *SVG and X3D* are fledgling graphics languages built on top of XML (see Chapter 5). They will enable XML (i.e. plain text) specification of vector graphics and 3D constructions, both of *huge* potential for GIS. See Chapter 4 for further discussion of SVG.

2.2.2 Separation of the user interface

The Web environment separates a user interface from the site of data processing. This separation poses problems for any operation that relies on a rapid response from the system. Consider the following examples.

1. A common GIS operation is to define a polygon using the "point and click" operation of the computer mouse. However, standard Web browsers treat a single mouse click as a prompt to transmit a query to the server. The delay between responses from the server is far too slow to maintain a coherent procedure.

2. Geographic queries are often context-sensitive. For instance a user may wish to have pop-up menus on a map that quickly display data about the properties of objects in space.

2.3 SOME EXAMPLES OF ONLINE GIS

Most of the existing online geographic information services fall into one of two categories: spatial query systems, and map-building programs. Many services incorporate both. As a result, most of the examples throughout this book are necessarily restricted to these functions, though there are some exceptions in Chapter 11. Only a few services provide more advanced GIS functionality, such as data analysis or modelling. Nevertheless, it is important to remember that this sort of functionality is also possible in the online environment. As we shall see in later chapters, the standards and tools needed to enable advanced GIS applications across the Internet are now currently under development. Many advanced applications are therefore likely to follow in coming years.

2.3.1 Spatial queries

The most widely seen GIS functions online are the queries based on spatial location. These include both selecting spatial objects, such as countries or states, and random points. Objects are most often selected by text interface (e.g. a list or database query form). Random points are usually selected via an *imagemap*, form image or java applet (see Chapter 4). Web browsers work on the hypertext concept of point and click. Few systems permit more sophisticated searches, such as those that require the user to select more than a single point, for example drawing a polygon to define a region.

Some early examples of services offering spatial queries include the following.

The Virtual Tourist

The Virtual Tourist (Plewe, 1997) was a service that aimed to provide tourists and the tourist industry with detailed information about every country in the world. Starting from a map of the world, users could click on successive maps until they had selected a country or region of their choice.

The VT was one of the earliest demonstrations of the power of online hypermedia to create systems that are more than the sum of their parts. The original service (which has since been commercialised), placed a central indexing site on top of literally thousands of contributing sites that provided sources of information about individual countries. A second, but equally important lesson was that it is easier to keep information up-to-date if detailed information is published online by the organisations that maintain it, rather than by a central agency that would be hard-pressed to keep the information current (see Chapter 6).

Pierce County Databases

In the state of Washington, USA, Pierce County provides a public GIS online. Called MAP-Your-Way!™, the system provides a flexible and widely accessible interface for many of the county's public databases (Fig. 2.1). Placing the system

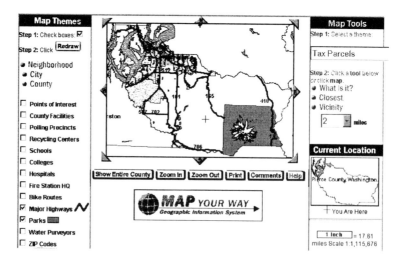

Figure 2.1. Online interface to Pierce County's MAP-Your-Way!™ information system.

online means that anyone can generate customised maps for any combination of features for any part of the county. The service was implemented using ESRI mapping tools.

The site includes a disclaimer that covers the following crucial matters: data limitations, interpretations, spatial accuracy, liability, and warranty against commercial use.

2.3.2 Map-building and delivery

Online facilities for building user-defined maps are now common. The following examples illustrate the range of features that online map-building services provide.

Xerox PARC World Map viewer

The first online map-building program was the Xerox PARC Map Viewer (Xerox 1993). The original system drew simple vector maps of any part of the world, at any resolution (Fig. 2.2). Originally the service was entirely menu-driven. That is users selected options to build new maps (e.g. zooming in by a factor of 2) one at a time from the options provided. More recent versions have added new features and enhanced the interface.

City street maps

In several countries, services now make street maps of various towns and cities available online. Some systems incorporate street maps with other data, such as telephone directories, or tourist information.

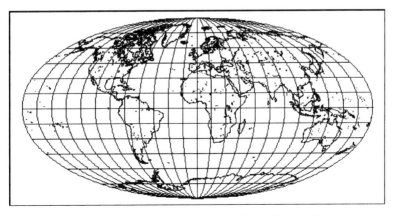

Figure 2.2. The Xerox PARC Map Viewer was the first online mapping system. Courtesy of Xerox, Palo Alto Research Center.

In some cases the maps are generated on demand, a simpler approach is to scan the maps to provide GIF images at various set resolutions. For instance, if a printed street directory consists of (say) 100 maps, then breaking each map into (say) 4×4 submaps, would require 1600 GIF images. To include the entire set at three resolutions (×1, ×2, ×4), would require 2100 images. Assuming that each image is no more than (say) 10 Kbytes, then the set can be stored just over 20 Mbytes. This volume is easily handled by any modern server. It has the advantage of eliminating the processing needed to draw each map. The main issue in such a system is to implement an indexing system that finds the desired map efficiently.

Environment Australia's interactive map builder

One of the frustrations of traditional paper maps is that the information a user requires may be spread across two or more sheets. This may be because the area overlaps two map sheets, or because the user wants to combine data not normally displayed together. GIS overcome this problem by allowing maps to be built on demand by interactively combining any layers that the user requests and for any selected area. However, the majority of potential map users do not have immediate access to a stand-alone GIS. Online map building services remove this final obstacle by placing online an interface to their map layers and drawing algorithms.

A good example of this sort of service is the environmental resources mapping system provided online by Environment Australia (ERIN, 1999). As well as being able to draw base maps on demand for any area, the system also allows users to select further information from a large range of environmental data layers.

CSU's Map Maker

Another issue for map users is to be able to plot their own data on a map. The Map Maker service (Steinke et al., 1996), developed by Charles Sturt University, addresses the need of field researchers (and others) to be able to produce publication quality plots of regions with their own sites or other locations added. First trialed in 1993, this service has been publicly available since 1995. The address for the service is

```
http://life.csu.edu.au/cgi-bin/gis/Map/
```

The base layers are generated by the GMT system, which is a freeware package of Generic Mapping Tools (Wessel and Smith, 1991, 1995).

The Map Maker was perhaps the first online service that allowed users to plot their own data on custom designed maps (Fig. 2.3). Users enter into a form field the set of coordinates and labels for points that they wish to include. The Map Maker also allows a limited degree of searching for major cities.

2.3.3 Other GIS functions

Spatial queries and simple map-building are by far the most common GIS functions available online. However, GIS are also used for a wide range of other operations. Perhaps the most common are spatial modelling and spatial data analysis. At the time of writing very few such services are available online. For example, online processing of satellite images have been trialed but are not generally available because they tend to require a great deal of computation.

The following examples give some idea of the sorts of interactive functions that can be, and are, online.

2.3.3.1 Environment Australia's species mapper

Environment Australia's service Species Mapper (ERIN 1995) provided a system for querying its database of species distributions and plotting them on maps. However, the service went further than that. The service also generated and plotted models predicting the potential geographic distribution of individual species. To achieve this (using the BIOCLIM algorithm), the server carried out the following sequences of steps (here slightly simplified) in real time:

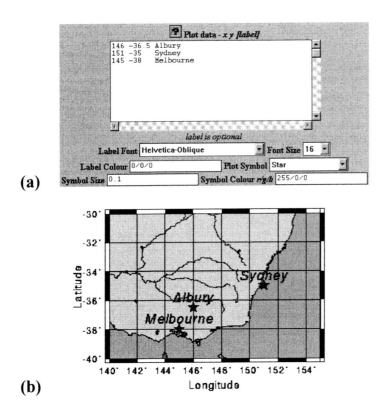

Figure 2.3. Adding user data to a map built online using CSU's Map Maker service. (a) Form for inserting user data. (b) The resulting map with the towns added.

1. Query database to retrieve records of species locations.
2. For each species location, interpolate values of essential climatic variables.
3. Calculate the climatic envelope bounding all the species records.
4. At the resolution specified, identify all other sites in the landscape that fall within the climatic envelope.
5. Plot the sites identified on a base map.
6. Deliver the map to the user.

Redevelopment of Environment Australia's databases to the service being taken off-line during 2000, but it is hoped that it will be reinstated in the future.

2.3.3.2 US Census maps

The TIGER Mapping System for USA (Figs 2.4, 2.5) is one of the most prominent online mapping facilities. It was established in 1994 and generates 45,000 to 50,000 maps per day. It aims

> "*to provide a good-quality, national scale, street-level map to users of the World Wide Web. This service is freely accessible to the public, and based on an open architecture that allows other Web developers and publishers to use public domain maps generated by this service in their own applications and documents.*" *(NGDC 2000)*

Future enhancement plans include street-level detail for the entire United States and more cartographic design features. Planned technical innovations include: open interface to images (through mapgen script), which will allow users to request maps directly to include in other documents, and reverse coordinate decoding, which will allow users to send a pixel coordinate (x,y) from the map and retrieve the associated real-world coordinate (thus making viable third party applications that rely on the interactive maps-provided by the service).

Two recent technologies also promise interesting new applications: satellite navigation systems (GPS) are readily available in cars, while internet access from wireless devices (mobile phones etc.) is on the rise. Thus we should see a whole suite of new synergetic applications (Chapter 11).

2.4 LARGE-SCALE SPATIAL DATA COLLECTION PROJECTS

2.4.1 The rise of online data warehouses

We saw above how the Virtual Tourist linked together information from many sources. However, that was a case of creating a unifying umbrella for online resources that had already appeared. In many other cases, the organisation and development of online data sources required a more systematic and coordinated approach.

One example was cooperation amongst sites dealing with related topics. This process led to the formation of information networks dealing with a wide range of topics (Green, 1995). This process still continues and has led to technical innovations to try to promote cross-site coordination. The first priority was to develop methods of indexing information. This goal has led to numerous indexing and production standards, such as XML, CORBA, and others that we shall discuss in later chapters.

There has also been a variety of software developments to enhance cross-site queries and information sharing. The first of these was the notion of a web crawler,

(a)

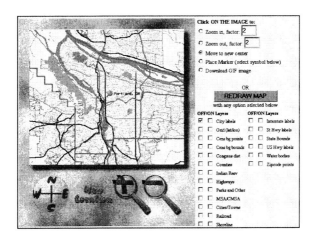

(b)

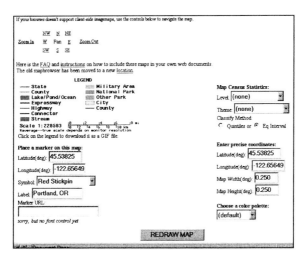

Figure 2.4.. Online user interface to the TIGER mapping system provided by the US Census Bureau. (a) The interacting map display, (b) Further form fields for customising outputs. Source: http://tiger.census.gov/

an automatic software agent that trawls the Web recording and indexing everything it finds. Many Internet search engines have used this kind of software. Later developments have included collaboration whereby a single query can spawn searches on many different sites, after which the results are pooled and transmitted back to the user.

However, the ultimate need is to develop systems that return not indexes of links to data, but the data itself. That is, they draw on data from many different sites and combine it into a single set of information for the user. This need gives rise to the idea of a *distributed data warehouse*. Data warehouses are simply assemblies of many different databases that are combined into a single information system. With the spread of electronic commerce and large-scale data collection systems, data warehouses are now common. A distributed data warehouse is just a data warehouse that is spread across several sites on the Internet. In some contexts (such as environmental information, biotechnology) there is a trend towards global information systems that integrate similar kinds of information worldwide. We will look at data warehouses more closely in Chapter 10.

It is this kind of information that is needed for online GIS. No single agency can store and maintain all information, in complete detail, for every part of the world. Perhaps the ultimate online GIS would be a global information system that integrated all kinds of geographical data from all sites into a single universal "atlas of the world". Unlike a traditional paper atlas, there is in principle no reason why

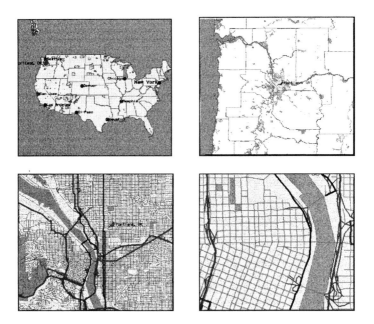

Figure 2.5. Zooming in with the TIGER Mapping System allows the user to select any desired view, from a continental scale to detailed street maps. In this online example, the view zooms in to downtown Portland, Oregon.

such a system needs to be restricted to a particular scale, or to particular data layers. In practice, however, there are severe technical and practical considerations that stand in the way.

2.4.2 The need for metadata

The desire to coordinate data across many different sites leads to the need for ways to label and identify data resources. *Metadata* are data about data. Metadata play a crucial role in recording, indexing and coordinating information resources (Cathro 1997).

Look at any book. On its cover you will nearly always find the name of the book, the name of its author and the name of the publisher. These pieces of information are metadata. Inside the front cover you will find more metadata: the copyright date, the ISBN, the publisher's address, and more. When you are reading the book, or if it is sitting on a bookshelf by your desk, then these metadata are rarely important. But if you want to find it in a large library, which might hold a million books, then you depend entirely on that metadata to find it. One of the worst things that can happen in a library is for a book to be put back in the wrong place. If that happens, then no one can find it without a long and painful search through the shelves.

As it is for books, so it is for online data. Metadata are crucially important for storing, indexing and retrieving items of information from online sources. Later in this book (especially Chapters 8 and 9), we will look in detail at the ways in which metadata are structured and used. In other chapters (6, 7 and 10), we will look at the ways in which metadata help to create large-scale information resources online.

Metadata have tended to be about content, creation and ownership as we shall see in Chapter 7. But online documents may require additional metadata, in terms of how and who shall granted online access.

2.5 DIFFERENCES BETWEEN STAND-ALONE AND ONLINE GIS

The traditional model for a GIS assumes that the system consists of a single software package, plus data, on a single machine. This model no longer meets the realities of many GIS projects, which today are often multi-agency, multi-disciplinary, multi-platform, and multi-software. Large numbers of contributors may be involved, and there may be a large pool of potential users. These users may require not only maps, but also many forms of multi-media output (e.g. documents). Moreover, they are likely to require access to the most current data available, and not to copies that may be months, or even years, old. A central practical issue, therefore, is how to provide widespread, device-independent, access to a GIS for large numbers of contributors and users.

The chief difference between an online GIS and traditional systems is the separation of user interface, data storage and processing. In stand-alone GIS all of these elements are normally present on a single machine. In online GIS the elements are normally spread across several machines.

This separation of GIS elements creates its own special issues and problems. One is the need to transfer data from one site to another. This leads to a need to cut down the number of transfers where possible. Another problem is that keeping track of what the user is currently doing becomes a non-trivial matter. Normally these details are immediately available because they are stored in computer memory while the GIS software is running. However, across a network they need to be stored so that the software can be initiated with the correct state.

In the remainder of this chapter, we look at some of the options for dealing with this separation of elements. In the chapters that follow, we look at some of the issues in the course of implementing GIS in an online environment.

2.6 OPTIONS FOR IMPLEMENTING ONLINE GIS

There are several ways in which GIS can be placed online. Perhaps the most telling is the way in which the user interface is handled. There are essentially two approaches. One is to take an existing stand-alone GIS and enable it to run online. The other is to provide GIS functionality via standard Web browsers. In the following sections we examine each of these options in turn. We then look at some of the issues that need to be considered.

2.6.1 Internet enable an existing stand-alone GIS

In this approach the user interface is a stand-alone GIS, which handles all (or most) of the processing involved. However, the system is provided with the capability to access files across the Internet. So local data (i.e. data residing on the user's machine) can be combined with data from other sites.

The above approach has several advantages:

- It retains all the power, speed and functionality of a traditional system.
- User's can continue to use a system that they are familiar with.

However, there are also some disadvantages:
- Users need to acquire and install specialist software (and often hardware), which may be expensive.
- The system is not universally available. Its audience is limited to those who have the necessary software.

2.6.2 GIS functionality in standard Web browsers

In this approach, a GIS is built to use existing standard browsers as its user interface. Normally this means that most of the data, and most of the processing, is delivered via a server, which then sends the results to the client for display.

The main advantages of this approach are:

- The system is potentially available to any web user (though browser add-ons may be required).
- Simple GIS can be implemented more quickly and easily than using the

power of a full system (why use a sledgehammer to crack a walnut?). For instance if all you want is for users to be able to make geographic queries of a database (e.g. "what sites fall within the selected region?"), then just a few geographic operations are required. The manager therefore need only install software to handle those operations. The rest of the processing can be handled with standard Web tools.

Some disadvantages are:

- The GIS manager needs to develop the GIS functionality. Few systems are available to pre-package the necessary storage, processing and display functions.
- Even if a standard GIS package is used as a backend, interfaces needs to be developed between it and the server and for converting outputs into a form that can be displayed on a browser.
- Many interactive processes, which are taken for granted on a stand-alone system, become difficult to sustain in a distributed environment.

Besides the above two extreme approaches there are various mixed alternatives. For example, an online GIS could use a standard Web browser for the starting interface, and for routine search and selection functions. However for specialised geographic operations and display it might pull up a GIS display package as a "helper" application. Some GIS vendors have developed technology for Web access.

2.6.3 Issues for implementing online systems

If existing GIS software is to be Internet enabled, then the Internet queries can be seen as akin to reading files in a traditional system, except that the files are stored somewhere else on the Internet, rather than on the user's hard disk. The rest of the system operates just as any stand-alone GIS would. Thus the main issue is how to build and send queries across the Internet, and how to handle the replies. We shall take up this issue in Chapter 5, where we address the issue of querying remote databases. Since this type of GIS applies essentially to existing software, it is mainly an issue for commercial developers.

Developing GIS that use the Web as their interface is a much more general concern. At the time of writing there are already many online GIS that use the Web as their interface. For the most part, these have been developed *de novo* by the sites providing the service. Very few pre-packaged toolkits are available for building online GIS. It is therefore important for potential developers to have a clear understanding of the issues involved. One of our aims in the following chapters is to help potential service developers by providing a set of tools to simplify the building of small online GIS.

In the following two chapters we shall examine issues and methods for implementing GIS that use the Web as the main platform. First we shall look at GIS processing on a Web server. We will then go on to look at processes that need to be handled by the Web client.

A fundamental practical issue for the development of online GIS is how it should relate to other services. Systems that are built in ad hoc and idiosyncratic fashion will not be consistent with other online GIS services. There are many advantages in developing GIS services that are consistent with those at other sites. As we shall see later (especially Chapter 7), standards are emerging that promote consistency, and provide a basis for Web mapping tools that simplify the development of GIS services online.

2.6.3.1 Connectionless interaction

Perhaps the most basic issues associated with HTTP1.0 are those arising from connectionless interaction. Web browsers and the HTML specification have been developed as a *stateless* interface into WWW resources. Under HTTP1.0, no context is preserved at either the client or server end, with each data transaction treated independently from any other. Later releases do allow preservation of context, but are not yet universal.

When interactive GIS software runs on a dedicated machine, there is an implicit assumption that the program's current state is a direct result of a user's interactions with it. Remote GIS cannot do this. Under the HTTP, for instance, every interaction of the user with the software is a fresh query and starts from the software's start up state. The combination of the separation of interface from processing, plus the connectionless nature of the communication, raises the need to provide mechanisms to maintain continuity.

There are several steps that can be taken to preserve context within a WWW service.

One method is to embed hidden state variables within Web documents. The HyperText Markup Language (HTML) provides several methods by which values for state variables can be transferred between server and client.

First, the HTML forms interface includes provision for "hidden fields" (see Section 3.3). We can use these to record a user's current "state" and essential portions of their interaction with the program. In effect, the server builds a script to reconstruct the current position as part of each interaction and embeds it in the document that it returns to the user. This information is not a simple log; previous panning and zooming can be ignored, for example.

A second method is the use of "cookies" (see Section 4.3). Amongst other things, cookies provide a way to identify users when they reconnect to a service. So, for instance, a user's preferences can be stored from session to session, thus allowing continuity.

A third procedure is caching. Creating a particular map, for instance, may require a series of operations that would soon grow impossibly tedious and time-consuming if repeated on each step. When interacting with a dedicated system we avoid this problem by saving the result of a series of steps. This is done either implicitly, as binary data linked with the user's session, or explicitly as a file under the user's name. In caching, the server not only delivers information back to the user, but also saves it to disk for a finite time. The file is then available as a starting point for the next query, if there is one. Caching requires careful documenting of the nature of each file: options include lookup tables, headers and coded file names. Cached files can be designated in many ways, including user (i.e. machine) and time. We have preferred designation by content as it avoids duplicating common operations for different users, and provides fastest response to common queries.

## CHAPTER 3

Server-side GIS operations

3.1 WEB SERVERS

In this chapter and the one that follows, we examine issues arising in the development and use of standard Web systems as a medium for GIS. Since we can run GIS packages, such as ArcInfo, over widespread client server systems such as X-Windows, we might ask what the Web protocol has to offer. And as we shall see in this chapter, Web based GIS is not without a few problems. Since the Web server's primary role is to deliver Web pages, server-side GIS operations have to be carried out by secondary programs (usually via the Common Gateway Interface, which we describe below). The original Web protocol, HTTP was designed for delivery of static text and simple images, and is not optimised for much else.

The gains come in low cost, package independence and in cross-platform web availability. But any program can package information up and transmit it according to HTTP. Some vendors are now doing this with their GIS packages. The other advantage is close integration with a web site which may have information and resources going way beyond the spatial.

An essential feature of geographic information systems is that they allow the user to interpret geographic data. That is, they incorporate features for data processing of various kinds. Some of these operations include: spatial database queries, map-building, geostatistical analysis, spatial modelling and interpolation.

In an online GIS, the question immediately arises as to where the above processing is carried out: at the *Web server*, which is remote from the user and supplies the information, or at the *Web client*, which receives the information and displays it for the user. In this chapter we look at issues and methods involved in carrying out GIS processing by a Web server.

What GIS operations can and should be run on a server? Only a few operations cannot be run on a server. These consist chiefly of interactive operations, such as drawing, that require rapid response to a user's inputs. Other operations, such as querying large or sensitive databases, must be carried out at the server. However, for most other operations, the question of whether to carry it out on the server, or on the client machine is not so clear cut.

The two most crucial questions when deciding whether an operation should be performed by the server or the client are:

❑ *Is the processing going to place too large a load on the server*? Any busy Web server will be accessing and transmitting several files per minute. The processing needed to (say) draw a map may take only a few seconds, but if requests for this operation are being received constantly, then they could quickly add up to an unmanageable load.

❑ *Does the volume of data to be sent to the client place too great a load on the network*? For instance, delivering large images constantly might slow response for the user.

3.1.1 The HTTP protocol

The "Hypertext Transfer Protocol" (HTTP) is the communications protocol used on the World Wide Web. It passes hypertext links from a browser to a server and allows the requested documents and images to be passed back to the browser. The server, or HTTP daemon (HTTPD) manages all communication between a client and the programs and data on the web site.

3.1.2 Hypermedia

The World Wide Web has turned the Internet into a medium for hypermedia. The term *hypermedia* is a contraction for hypertext and multimedia. Hypertext refers to text that provides links to other material. *Multimedia* refers to information that combines elements of several media, such as text, images, sound, and animation. *Hypertext* is text that is arranged in non-linear fashion. Traditional, printed text is linear: you start at the beginning and read through all the passages in a set order. In contrast, hypertext can provide many different pathways through the material. In general we can say that hypertext consists of a set of linked "objects" (text or images). The links define pathways from one text object to another.

Electronic information systems have led to a convergence of what were formerly very different media into a single form, known as "multimedia". Film and recordings, for instance, have now become video and audio elements of multimedia publications. In multimedia publications, one or another kind of element tends to dominate. In traditional publications, text tends to be the dominant element, with images provided to "illustrate" the written account. One exception is the comic book in which a series of cartoon pictures provides the storyline, with text provided as support.

Vision in humans is the dominant sense, so in multimedia visual elements (particularly video or animation) often dominate. However, in on-line publications, the speed of the network transmission is still a major consideration for narrowband delivery. At present full-screen video is not practical, except across the very fastest network connections.

It was the introduction of browsers with multimedia capability that made the World Wide Web a success. However, audio and video elements are not currently supported by any Web browsers. These require supporting applications that the browser launches when required. Examples include the programs *mpegplay* for MPEG video files, and *showaudio* for sound. Each kind of information requires its own software for generating and editing the material.

In this chapter we make occasional use of markup formats, distinguished by tags : start tags begin and end with angle brackets as <map>; end tags have an additional / as in </map>. The full syntax and structure of markup for HTML and XML are covered in Chapter 5.

3.1.3 Server software

There are many versions of software for Web servers. The most widely used is the Apache server, which is a freeware program, with versions available for all Unix operating systems. However, most of the major software houses have also developed server software.

It is not feasible here to give a full account of server software and the issues involved in selecting, installing and maintaining it. Here we can only identify some of the major issues that Web managers need to be aware of.

Security is a major concern for any server. Most packages make provision for restricting access to material via authorisation (access from privileged sites), and authentication (user name and password). With heightened concern over the security of commercial operations, most server software now includes provision for encryption as well as other features to minimise the possibility of illegal access to sensitive information.

In terms of functionality, most server software today makes provision for standard services such as initiating and running external programs, handling cookies (see Chapter 4), allowing file uploads and making external referrals. Perhaps the most important questions regarding functionality are what versions of HTTP the server software is designed for and how easily upgrades can be obtained and installed.

When installing a server, it is important to give careful attention to the structure of the two directory hierarchies. Source material used by HTTP servers normally falls into two main hierarchies:

❏ *The document* hierarchy, which contains all files that are delivered direct to the client.

❏ The CGI hierarchy (see next section), which contains all the files and source data needed in processing information.

The above distinction is important because it separates freely accessible material, such as documents, from programs and other resources used in processing, which often has security implications. Care is needed too in the organisation of directories under each of these hierarchies. The names of directories normally form part of the URL for any item of information, so it is important that they have logical names. Moreover, they should reflect the sorts of queries that users will make. Many Web managers make the mistake of structuring directories and information in terms of system management, or in terms of the internal organisation of their institution or corporation. So, for example, users would normally prefer to look for tourist information organised under country or region, rather than (say) the names of individual travel companies or hotel chains.

More importantly, the logical names of directories should never change once they are established. It is possible to circumvent this issue to some extent by using aliases to provide a logical hierarchy. In Unix, for instance, directories and files can be assigned logical names that are completely independent of their true storage location. For instance, the real file path

```
/documents/internal-data/file023.dat
```

might be assigned the much simpler logical path

```
hotel-list.
```

3.1.4 Practical issues

Although not specifically related to online GIS, a number of general issues are important for any online information service.

In maintaining a server, three important issues are system updates, backups and server logs. System updates consist of files and data that need to be changed at regular intervals. For example, a data file that is derived from another source may need to be downloaded at regular intervals. These sorts of updates can be automated by using an appropriate system software. In the Unix operating system, for instance, the traditional method of automating updates is by setting *crontabs* (a system for setting automated actions against times) to run the necessary shell scripts at regular intervals.

Backups are copies of data on a server. Their function is to ensure that vital data is not lost in case of hardware or other failure. Backups are usually made regularly on tapes. For safety, in case of fire, backup copies are best stored off site. Mirror images of data provide another form of backup. However, it may be unwise to rely on an outside organisation to provide the sole form of backup. As with server updates, backups can be automated. Any busy server would need daily backups, though if storage space is limited, these could be confined to copies of new or altered files. However, it is always wise to make a complete copy of the document and CGI hierarchies at regular intervals.

Server log files provide important sources of information about usage of the system. The *access log* lists every call to the system. As well as recording the names of files or processes that are accessed, it also includes the time and address of the user. This information can be useful when trying to assess usage rates and patterns. The *error log* is useful in identifying faults in information services, as well as potential attempts to breach system security.

3.2 SERVER PROCESSING

To be a complete range of media, the Web provides a methodology by which a Web server can carry out processing of various kinds in order to respond to a query. The processing falls into four main classes, which are listed below.

- ❑ Allowing users to submit data to the server, usually via forms.
- ❑ Uploading files to the server.
- ❑ Data processing to derive the information required to answer a query.
- ❑ Building and formatting documents and their elements.

The *Common Gateway Interface* (CGI) is a link between the Web server and processes that run on the host machine (Fig. 3.1). CGI programs are accessed through the web just as a normal HTML document is, through a URL. The only condition placed upon CGI programs is that they reside somewhere below the cgi-bin directory configured within the HTTPD server. This is so the server knows whether to return the file as a document, or execute it as a CGI program.

The data from a form or query is passed from the client browser to the HTTPD server:

1. The HTTPD server forwards the information from the browser via CGI to the appropriate application program.
2. The program processes the input and may require access to certain data files residing on the server, such as a database.
3. The program writes either of the following to standard output: a HTML document; or a pointer to a HTML document in the form of a Location Header that contains the URL of the document.
4. The HTTPD server passes the output from the CGI process back to the client browser as the result of the form or query submission.

Programs to run CGI processes can be written in virtually any programming language. The most common ones being PERL, C and shell scripts. The basic structure of a CGI program is illustrated by Figure 3.2.

PERL is the language of choice for many CGI programmers due to its powerful string manipulation and regular expression matching functionality. This makes it well suited to handling both CGI input from HTML forms, as well as producing dynamic HTML documents.

3.2.1 CGI and form handling

3.2.1.1 CGI Input

HTML forms are implemented in a way that produces key-value pairs for each of the input variables. These key-value pairs can be passed to the CGI program using

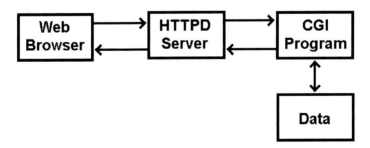

Figure 3.1. Role of the Common Gateway Interface (CGI), which mediates exchanges between a web server and other programs and data.

one of two methods, GET and POST. An important point to note is that with both methods, the data is encoded so as to remove spaces, and other various characters from the data stream. A CGI program must decode the incoming data before processing.

3.2.1.2 GET

The GET method appends the key-value pairs to the URL. A question mark separates the URL proper from the parameters which are extracted by the HTTPD server and passed to the CGI program via the QUERY_STRING environment variable. The general format of a GET query is as follows:

```
http://server-url/path?query-string
```

In this syntax, the main terms are as follows:

server-url	is the address of the server that receives the input string;
path	is the name and location of the software on the server;
query-string	is data to be sent to the server.

CGI Program Structure

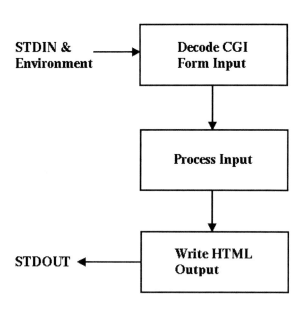

Figure 3.2. The structure of CGI programs. Any program must carry out the three tasks shown: first it decodes any form input, then carry out the required processing; and finally it must compile an output document to return to the user.

Below are some typical examples,

```
http://www.cityofdunedin.com/city/?page=searchtools_street
http://www.geo.ed.ac.uk/scotgaz/scotland.imagemap?306,294
http://www.linz.govt.nz/cgi-bin/place?P=13106
http://ukcc.uky.edu:80/~atlas/kyatlas?name=Main+&county=21011
```

The GET method is generally used for processes where the amount of information to be passed to the process is relatively small, as in the above examples. Where larger amounts of data need to be transmitted, the POST method is used.

3.2.1.3 POST

In the POST method, the browser packs up the data to be passed to the Web server as a sequence of key-value pairs. When the server receives the data it passes it on to the CGI program as standard input. Here is a typical string that would be sent by the sample form, which is described in the section that follows.

```
register=tourism&country=Canada&attraction=Lake%Louise&
description=&latdeg=51&latmin=26&longdeg=-116&longmin=11&
region=Alberta&hotel=YES&meals=YES&park=YES&history=YES
&website=http://www.banfflakelouise.com/&
email=info@banfflakelouise.com
```

The application program needs to unpack this string and carve it up into the required name-value pairs before it can use the data.

3.2.2 Forms and image fields

An important aspect of the Web is that information access need not be passive. Users can send data to Web servers via forms. Forms are documents that include fields where data can be entered. These fields consist of text boxes and various other "widgets" (Table 3.1). Users normally submit form data by clicking on a SUBMIT button. The server receives the submitted data, processes it and sends a response back to the user (Fig. 3.2).

The basic structure of an HTML form is as as follows:

```
<form processing_options>
        input fields mixed with text
</form>
```

Here the "processing options" are attributes that describe the method that will be used to transmit the data (see GET or POST above), and indicate which program will receive and process the form data. The types of input fields that can be included are listed in Table 3.1.

Table 3.1. Some common widgets in HTML forms.

Field	*HTML source code*	*Resulting widget*
Text	`<input name="field1" type="text" size="20" value="Sample text">`	Sample text
Textarea	`<textarea name="message" rows="2" cols="15"> </textarea>`	Sample text
Select	`<select name="country">` `<option value="UK" >Britain` `<option value="CAN">Canada` `<option value="USA">USA` `</select>`	Britain ▼ Britain Canada USA
Radio	`Size <input type="radio" name="size" value="100">100` `<input type="radio" name="size" value="1000">1000`	Size ○ 100 ○ 1000
Check box	`Capital city? <input type="checkbox" name="capital" value="TRUE">`	Capital city? ☐
Hidden field	`<input type="hidden" name="scale" value="100">`	Not visible
Reset	`<input type="reset" value="Clear Fields">`	Clear Fields
Submit	`<input type="submit" value="Submit entry">`	Submit entry

Example

Below is the HTML source code for a simple form that might be used for (say) operators to register tourist attractions in an online database. The resulting Web form as it would appear in a browser is shown in Figure 3.3.

The code contains two hidden fields, which provide technical data to the server. One, called "register", tells the server which register to use. This is essential if the same software handles several different services. The second hidden field tells the server which country the registered sites are located in (Canada in this case). This data would be necessary if the underlying database held information for many countries, but the form applied to one only.

Note the use of the table syntax to arrange a clear layout of the fields. Web authoring systems usually make it possible to construct HTML documents and forms without seeing the underlying code at all. However, it pays authors to learn to manipulate HTML code directly. For instance, many automatic form builders use features that may not display well on all browsers.

```
<html>
<body>
<h1>Tourist site register</h1>
```

```
<form action="http://life.csu.edu.au/cgi-bin/geo/demo.pl"
    method="POST">

<input type="hidden" name="register" value="tourism">
<input type="hidden" name="country"  value="Canada">

<p>Attraction
<input name="attraction" type="text" size="50" value="Enter
its name here">

<p>Description
<textarea name="description" rows="2" cols="40">
Write a brief description here.
</textarea>

<h4>Location</h4>
<table>
<tr><td><i>Latitude</i><br>
    <input name="latdeg"  type="text" size="3"> deg
    <input name="latmin"  type="text" size="3"> min
   
<td><i>Longitude</i> <br>
    <input name="longdeg" type="text" size="3"> deg
    <input name="longmin" type="text" size="3"> min

<td><i>Province</i> <br>
<select name="region">
<option value="BC">British Columbia
<option value="ALB">Alberta
<option value="SAS">Saskatchewan
<option value="MAN">Manitoba
<option value="ONT">Ontario
<option value="QUE">Quebec
<option value="NB">New Brunswick
<option value="PEI">Prince Edward Island
<option value="NS">Nova Scotia
<option value="NFL">Newfoundland
</select>


</table>

<h4>Facilities available</h4>

<table>
<tr><td valign="top"><i>Hotel</i> <td>
<td><input type="radio" name="hotel" value="YES"> Yes
<br><input type="radio" name="hotel" value="NO" > No
   
```

```
<td valign="top"><i>Meals</i>
<td><input type="radio" name="meals" value="YES"> Yes
<br><input type="radio" name="meals" value="NO" > No
</table>

<h4>Features</h4>

Park    <input type="checkbox" name="park"    value="TRUE">
History <input type="checkbox" name="history" value="TRUE">

<h4>Online addresses</h4>

<table>
<tr><td><i>Website</i> <br>
<input name="website" type="text" size="30" value="http://">

<td><i>Email</i> <br>
<input name="email" type="text" size="20" value="Enter
name@location">
</table>
<input type="reset"  value="Clear">
<input type="submit" value="Submit">

</form>
</body>
</html>
```

Image input fields

HTML provides online forms with the capability for images to be used as input fields. That is, if a user points at an image with a mouse and clicks the mouse button, then the location of the mouse pointer will be transmitted as image coordinates. This facility makes it possible to provide maps, within a form, that allow the user to select and submit geographic locations interactively.

   HTML provides for a user defined widget called an *image field*. Although we restrict the discussion here to images that are maps, the image can potentially be anything at all. A typical entry for an image field might be as follows.

```
<input type="image" name="coord" src="world.gif">
```

In this example, the name of the field variable that is defined is **coord** and the image to be displayed is contained in the file called **world.gif**. The image field so defined has the following important properties.

❑   The browser transmits not one, but two values from this field. They are the x and y coordinates from the image, and are denoted by **coord.x** and **coord.y** respectively.

❑ These coordinates are image coordinates, not geographic coordinates. It is up to the processing software to make the conversion to latitude and longitude.

❑ The image field acts as a SUBMIT button. That is, when the user clicks on the image the form data (including the image coordinates) are submitted to the server immediately.

To convert the image coordinates into geographic coordinates, we need to know the size of the image in pixels, and the Latitude and Longitude corresponding to the two corners. The simplest formula for converting the image coordinate `coord.x` into the corresponding longitude L is then

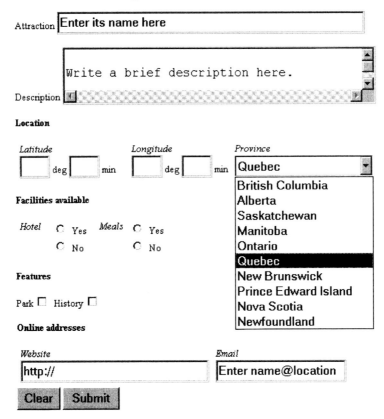

Figure 3.3. Example of a World Wide Web form. This form is defined by the HTML source code presented in the text. The menu for province selection has been pulled down and the entry for Quebec highlighted.

where L_0 and L_1 and the longitudes represented by the top and right sides of the map and x_{max} is the horizontal width of the image in pixels. A similar formula would apply for extracting latitude from `coord.y`. Note that these formulae apply only on small scales. On a global scale it is necessary to take into account the map projection that is used.

3.2.3 Quality assurance and forms

The distributed nature of the Web means that hundreds or even thousands of individuals could potentially contribute data to a single site. This prospect raises the need to standardise inputs as much as possible and to guard against errors. One method, which can be used wherever the range of possible inputs is limited, is to supply the values for all alternatives. So, for instance, instead of inviting users to type in the text for (say) "New South Wales" (ie. entering it as a text field), we can supply the name as one of several options, and record the result in the desired format (e.g. "NSW"), by using a pull-down menu (as shown for the field `province` in Figure 3.3). This method avoids having to sort out the many different ways in which the name of the state could be written (e.g. "NSW", "N.S.W.").

3.3.4 Processing scripts and tools

The most basic operation on a server is to return a document to the user. However, many server operations need to include some form of processing as well. For instance, when a user submits a form the server usually needs to interpret the data in the form and do something with the data (e.g. write it to a file, carry out a search), then write the results into a document that it can return to the user. The processing may include passing the data to various third party programs, such as databases, or mapping packages. Some commercial publishing packages now provide facilities to install and manage the entire business. However, the processing itself is often managed using processing scripts. Scripts are short programs that are interpreted by the system on the fly. They are usually written in scripting languages, such as Perl, Python, Java (see Chapter 4), Shell Script (Unix/Linux), or Visual Basic (Windows).

  The following short example shows a simple Perl script that processes form data submitted to a Web server. The code consists of three parts. Part 1 decodes the form data (which is transmitted as an encoded string), and the data are stored in an array called **list**. This array, which is of a type known as an *associative array*, uses the names of the fields to index the values entered. So for the field called **country** in the form example earlier (which took the value Canada), we would have an array entry as follows:

```
list{country}=Canada .
```

  Part 2 writes the data in SGML format (see Chapter 5) to a file on the server. Part 3 writes a simple document (which echoes the submitted values) to acknowledge receipt to the user. A detailed explanation of the syntax is beyond the scope of our discussion. For a detailed introduction to the freeware language Perl see Schwartz (1993). There are also many online libraries of useful scripts, tools, and tutorials.

```perl
#!usr/local/bin/perl
#   Simple form interpreter
#   Author:  David G. Green, Charles Sturt University
#   Date   :  21/12/1994
#   Copyright 1994 David G. Green
#   Warning: This is a prototype of limited functionality.
#            Use at your own risk. No liability will be
#            accepted for errors/inappropriate use.
#
# PART 1 - Convert and store the form data
        # Create an associative list to store the data
%list = ();
        # Read the form input string from standard input
$command_string = <STDIN>;
chop($command_string);
        # Convert codes back to original format
        # ... pluses to spaces
$command_string =~ s/\+/ /g;
        # ... HEX to alphanumeric
$command_string =~ s/%(..)/pack("c",hex($1))/ge;
        # now identify the terms in the input string
$no_of_terms = split(/&/,$command_string);
@word = @_;
        # Separate and store field values, indexed by names
for ($ii=0; $ii<$no_of_terms; $ii++)
{       @xxx = split(/=/,$word[$ii]);
        $list{$xxx[0]} = $xxx[1];
}
#
# PART 2 - Print the fields to a file in SGML format
$target_name = "formdata.sgl";
open(TARGET,">>$target_name");
        # Use the tag <record> as a record delimiter
print TARGET "<record>\n";
        # Cycle through all the fields
        # Print format <fieldname>value</fieldname>
foreach $aaa (keys(%list))
{       print TARGET "<$aaa>$list{$aaa}<\/$aaa>\n";
}
close(TARGET);
print TARGET "<\/record>\n";
#
# PART 3 - Send a reply to the user

        # Writes output to standard output
        # The next line ensures that output is treated as HTML
print "Content-type: text/html\n\n";
        # The following lines hard code an HTML document
```

```
print "<HTML>\n<HEAD>\n Form data
return\n<\/HEAD>\n<BODY>\n";
print "<H1>Form received.</H1>\n<P>Here is the data you
entered ...\n";
        # Print the fields in the form FIELD = VALUE
foreach $aaa (keys(%list))
{       print "Field $aaa = $list{$aaa}\n";
}
print "<\/BODY><\/HTML>\n";
```

To understand how this script would be used in practice, suppose that it is stored in an executable file named simple.pl that is located in the cgi-bin hierarchy of a server whose address is mapmoney.com. Then the script would be called placing the following action command in the form:

```
<form action="http://mapmoney.com/cgi-bin/simple.pl"
        method="POST">
```

The purpose of the above example is to show the exact code that can be used to process a form. However, in general, it is not good practice to write scripts that hard-wire details such as the name of the storage file, or the text to be used in the acknowledgment. Instead the script can be made much more widely useful by reading in these details from a file. For instance, the return document can be built by taking a document template and substituting details supplied with the form for the blanks left in the template, or by the script itself. Likewise, the name of the output file could be supplied as a run-time argument to the script. To do this the script would need to replace the line

```
$target_name = "formdata.sgl";
```

with an assignment such as the following

```
$target_name = @ARGV;
```

The URL to call the script would use a "?" to indicate a run-time argument:

```
http://mapmoney.com/cgi-bin/simple.pl?formdata.sgl
```

This example still has problems. In particular, this example shows the extensions of both the script and the exact name of the storage file. For security reasons, it is advisable to avoid showing too many details.

3.3 ONLINE MAP BUILDING

To build a simple map across the Web, the following sequence of steps need to take place:

1. the user needs to select or specify the details of the map, such as the limits of its borders and the projection to be used;
2. the user's browser (the "client") needs to transmit these details to the server;
3. the server needs to interpret the request;
4. the server needs to access the relevant geographic data;
5. the server needs to build a map and turn it into an image (e.g. GIF format);
6. the server needs to build an HTML document and embed the above image in it;
7. the server needs to return the above document and image to the client;
8. the browser needs to display the document and image for the user.

This process is illustrated in Figure 3.4. In the above sequence, only Steps 2, 7, and 8 are standard operations. The rest need to be defined. In almost all cases, Step 1 involves the use of a form, which the browser encodes and transmits. In Step 3, the server passes the form data to an application program, which must interpret the form data. The same program must also manage the next three steps: communicating with the geographic data (Step 4), arranging the map-building (Step 5) and creating a document to return to the user (Step 6).

When building a map in the above example, what the system actually produces is a text document (which includes form fields) with the map inserted. The map itself is returned as a bit-map (pixel-based) image. The image is in some format that a Web browser can display. Until recently, this usually meant a GIF format, or even JPEG. Having to convert Vector GIS data to a pixel image has

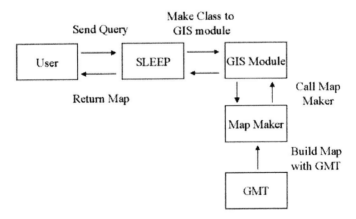

Figure 3.4. The flow of information from user to server and back that is involved in a typical system for building maps over the Web. The modules used here are as follows: SLEEP is a script interpreter (see Section 3.5) with a module of GIS calls to the Mapmaker software (Steinke et al. 1996), which uses the GMT freeware package of map-drawing tools (Wessel and Smith 1991, 1995).

been a severe drawback. The output loses precision. It cannot be scaled. And downloading a pixel image, even a compressed one, often requires an order of magnitude more bandwidth than the original data. A solution is now at hand with the introduction of a standard for Scalable Vector Graphics (SVG). We describe this new standard in the next chapter (Section 4.4.5).

3.4 THE USE OF HIGH-LEVEL SCRIPTING LANGUAGES

Programming languages were invented to simplify the task of programming computers. High-level languages are computer languages that are designed to simplify programming in a particular context. It is far easier to write a program that carries out a specialised task if you can use terms and concepts that relate directly to the system concerned. The problem with general purpose languages is that the solution to a programming problem has to be expressed in terms that are very far removed from the problem's context. In particular, most automating of online services has involved writing programs in languages such as Perl, Java, C++ or shell script.

It has become commonplace in many computing packages to simplify the specification of processing steps by providing high level scripting languages. The advantages are that most operations can be programmed far more concisely than general purpose languages. Also, because they are oriented towards a specific content area they are usually easier to learn and to use. For example to extract the contents of a Web form in the language Perl requires a program of at least a dozen lines. However, in a Web publishing language the entire process is encapsulated in a single command.

High-level languages are desirable in developing server-side operations on the Web. The advantages (Green 1996, 2000, Green et al. 1998) include modularity, reusability, and efficiency. As we shall see below, and in later chapters, high level languages can include GIS functions and operations.

Most conventional GIS systems incorporate scripting languages to allow processes to be automated. In automating a web site, particular scripts can be generalised to turn them into general purpose functions. To do this we start with a working script such as the following simple example.

```
SET BOUNDS 34.8S 140.1E 40.4S 145.2E
EXTRACT  roads, topography, vegetation
PLOT roads
PLOT topography
PLOT vegetation
```

We then replace constant values by variables, which can be denoted by angle brackets.

```
SET BOUNDS <tlat> <tlong> <blat> <blong>
EXTRACT  roads, topography, vegetation
PLOT roads
PLOT topography
PLOT vegetation
```

This generalised script can now serve as a template for producing a plot of the same kind within any region that we care to select. If the user provides the boundaries from (say) a form, then we could generate a new script by using a Perl script to replace the variables in the template with the new values. Here's a simple example of a Perl script that does this.

```
#! /usr/bin/perl
# Build a simple script from a template
# The associative array markup contains
replacement values
getvarsfromform;
filtertemplate;

sub filtertemplate {
        while ($sourceline=<STDIN>)
        {       chop($sourceline);
                $targetline = $sourceline;
        # Enter the input string into the
template fields
                for ($i=0; $i<$no_of_tags; $i++)
                {       $work = $tag[$i];
                        $targetline =~
s/$work/$formvar{$work}/gi;
                }
                print "$targetline\n";
        }
}
```

This script acts as a filter. Its function is similar to a merge operation in a word processor. We supply values for variables in a form. The function **getvarsfromform** (cf. the example in Section 3.3.4) retrieves these values as a table ($formvar). The perl script then reads in the template as a filter and prints out the resulting script. To run this script we would use a call such as

```
cat  templatefile | filterfile  >  outputscript
```

where **templatefile** is the file containing the template, **filterfile** is the file containing the above perl code and **outputscript** is the resulting script.

Although the above procedure works fine, it is cumbersome to have to rewrite perl scripts for each new application. A more robust and efficient approach is to continue the generalisation process to include the perl scripts themselves. This idea leads quickly to the notion of implementing web operations via a high-level publishing language.

The following example of output code shows what form data might look like after processing by a script such as the above. The format used here is XML (see Chapter 5), with tags such as <country> corresponding to form fields.

```
<country>
<name>United State of America</name>
<info>http://www.usia.gov/usa/usa.htm/</info>
<www>http://vlib.stanford.edu/Servers.html</www>
<government>http://www.fie.com/www/us_gov.htm</government>
<chiefs>http://www.whitehouse.gov/WH/html/handbook.html
</chiefs>
<flag>http://www.worldofflags.com/</flag>
<map>http://www.vtourist.com/webmap/na.htm</map>
<spdom>
        <bounding>
              <northbc>49</northbc>
              <southbc>25</southbc>
              <eastbc>-68</eastbc>
              <westbc>-125</westbc>
        </bounding>
</spdom>
<tourist>http://www.vtourist.com/webmap/na.htm</tourist>
<cities>http://city.net/countries/united_states/</cities>
<facts>http://www.odci.gov/cia/publications/factbook/us.html<
/facts>
<weather>http://www.awc-kc.noaa.gov/</weather>
<creator>David G. Green</creator>
<cid>na</cid>
<cdate>1-07-1998</cdate>
</country>
```

The following simple publishing script converts the above data from XML format (held in the file **usa.xml**) into an HTML document (stored in the file **usa.htm**). In this case it simply replaces the tags with appropriate code. Such a script can be easily prepared by a naïve user without understanding what the conversion operations actually are.

```
var <country>          </head><body bgcolor="#ffffff">
...... definitions omitted ......
source  usa.xml
target  usa.htm
new all
sub
close all
```

3.5 IMPLEMENTING GEOGRAPHIC QUERIES

3.5.1 Example –Great Circle Distances

A simple example of a geographic query online is the following demonstration of computing great circle distances between points selected from a map of the world. The interface (Fig. 3.5) is a form with a map image in it.

To call the example, the URL addresses the interpreter for this service (here it is called **mapscript**). The interpreter links to a set of functions that perform relevant GIS functions. To enable the demonstration, we pass to the interpreter the name of the publishing script to be used (here it is in the file **circle.0**). The full call is therefore as follows:

```
http://life.csu.edu.au/cgi-bin/gis/demos/mapscript?circle.0
```

The complete process needs two publishing scripts. The script above (mapscript) provides an entry into the service (Fig. 3.6). It defines initial locations for the two points involved. A second script (circle.1) processes subsequent calls made from the form. The complete list of files involved is thus as follows:

- ❑ An initialisation script (circle.0);
- ❑ A processing script (circle.1);
- ❑ The document template (circle.xml), essentially it is the form shown in Fig. 3.5, but with variables in place of the locations of the two points;
- ❑ The map used, saved as a static GIF image (world.gif);
- ❑ The interpreter (mapscript).

The HTML file that is displayed on the client browser does not exist as a stored file. It is generated on the fly by the server (see Fig. 3.7). The key element in the form is the image input type, given by the element

```
<input type="image" name="coord" src="world.gif>
```

Great circle distances

This service calculates the distance from the point A to point B.
Click on the image to select a new point.
Lat/Longs are in degrees North.
Current distance `732.9` Km

Value	Point A	Point B	
SELECT POINT		●	Submit Query
Latitude (e.g. -34.5)	-37.45	-33.53	Reset
Longitude (e.g. 134.5)	144.58	151.1	

Figure 3.5. A simple form interface for computing great circle distances. The user simply clicks on the map to select a new location for point A or point B. A radio button (SELECT POINT) defines which point is being selected. Following each selection, the server regenerates the form with the new settings. The great circle distance is displayed in the box at the right of the map.

```
source circle.xml      (define the template file)
target STDOUT          (send results to standard output)
var alat   -34.5       (initial Latitude of Point A)
var along 147          (initial Longitude of Point A)
var blat   -33         (initial Latitude of Point B)
var blong 149          (initial Longitude of Point B)
radius      6306       (Earth's radius in kilometres)
form                   (extract fields from the form)
circle                 (calculate the distance)
sub                    (place new values in the template)
```

Figure 3.6. The publishing script used in the great circle example.

Clicking on the image generates a pair of values, here called **coord.x** and **coord.y**, which are the coordinates of the point on the image where the mouse is clicked.

Below is the HTML source code for the form used in the great circle example. The lines in bold are (in order): (i) the call to the form processing script; (ii) an image input field, which ensures that coordinates are read off the map; (iii) a hidden field providing a value for the earth's radius; (iv) the final four lines in bold are fields that were processed by the publishing script, as shown in Fig. 3.6.


```
<html>
<head>
<title>Great circle distance calculations</title>
</head>
<body>
<form
action="http://life.csu.edu.au/cgi-bin/gis/calc?circle.1"
                     method="POST">                        (i)
<table>
<tr><td><input type="image" name="coord"
       src="http://life.csu.edu.au/ gis/demos/world.gif">
                                                           (ii)
<td><h2>Great circle distances</h2>
This service calculates the distance from the
point A to point B.
<br><i>Click on the image to select a new point. </i>
<br>Lat/Longs are in degrees North.
<br><b>Current distance
<input type="text" name="circle"
       value="732.9" size="10"> Km
<input type="hidden" name="radius" value="6366.19">
       (iii)
</b>
```

```
</table>
<p>
<table border="1" cellpadding="1">
<tr><td><b>Value</b> <td><b>Point A</b>
<td><b>Point B</b>
<td rowspan="2"><input type="submit">
<tr><td>SELECT POINT
<td><input type="radio" name="site" value="source" >
<td><input type="radio" name="site" value="target"  checked>
<tr><td>Latitude (e.g. -34.5)
<td><input type="text" name="alat" value="-37.45">            (iv)
<td><input type="text" name="blat" value="-33.53">            (iv)
<td rowspan="2"><input type="reset">
<tr><td>Longitude (e.g. 134.5)
<td><input type="text" name="along" value="144.5">            (iv)
<td><input type="text" name="blong" value="151.1">            (iv)
</table>
</form>
</body></html>
```

3.6. CONCLUDING REMARKS

The above examples are indicative of the kind of scripting that developers of online GIS are likely to encounter. Most commercial systems have in-built scripting features, which simplify the implementation of GIS online, much as we have demonstrated above.

Rapid developments are presently under way in the technology available for

XML Template (input circle.xml)	HTML Form (standard output)

`<input type="text"`
`      name="alat"`
`      value="<alat>">`
	`<input type="text"`
	`      name="alat"`
	`      value="-34.5" >`

`<input type="text"`
`      name="blat"`
`      value="<blat>" >`
	`<input type="text"`
	`      name="blat"`
	`      value="-33">`

`<input type="text"`
`      name="along"`
`      value="<along>">`
	`<input type="text"`
	`      name="along"`
	`      value="147">`

`<input type="text"`
`      name="blong"`
`      value="<blong>"`
	`<input type="text"`
	`      name="blong"`
	`      value="149">`

Figure 3.7. Use of a template to update the form used in the great circle example. The publishing script combines input from the form (Fig. 3.5) takes and combines it with a template (above left) by replacing actual values (e.g. 147) for the corresponding XML variables (e.g. <along>). The resulting code for the new form is shown at right.

server-side processing. Within the next few years, these developments have the potential to transform the way in which online GIS is implemented and practised.

Some of the changes are likely to occur through innovations from the World Wide Web Consortium (W3C). One such development, which will be of considerable use, is the Scalable Vector Graphics (SVG) standard (see Chapter 4). Vector data usually requires far less storage space, and, more importantly, far less transmission bandwidth. Although the web has had image formats available since its inception, vector systems have typically been proprietary and activated via plug-ins. One of the most successful has been Flash from Macromedia. Now, we have a web standard for expressing vector operations in XML, which will enable CGI scripts to generate vector diagrams on the fly.

Another major development from W3C for server side operations is the update to the HTTP protocol, which will allow maintenance of state. Security mechanisms are getting more sophisticated, but, otherwise, there is not much to report server-side.

Another series of important initiatives are those being developed by the Open GIS Consortium (OpenGIS 2001). One proposal is the Geographic Markup Language, which we will look at more closely in Chapter 5. However, perhaps the most telling proposal is the specification that the Consortium is developing for Open GIS. We look at the Open GIS specifications in Chapter 7 and Chapter 10. One important development is the Web Map Server Interface (OpenGIS, 2001), which specifies the types of outputs that a map server should be capable of providing.

## CHAPTER 4

Client-side GIS operations

4.1 INTRODUCTION

One of the fundamental differences between online and stand-alone GIS is the separation of the user interface from the main data and processing, as we saw in Chapters 2 and 3. To have to refer every operation, every choice back to the server is slow. It is also frustrating for the user. Therefore any steps that are used to move interactions to the client's machine is desirable. Many types of operations can be performed client-side. The following list is indicative, but not exhaustive.

❑ Simple selection of geographic objects can often be implemented as client-side image maps.
❑ Some simple functions, such as elementary checks of validity for data entries, can be implemented in Javascript, now standardised as ECMA script, which comes as built-in to most Web browsers.
❑ Vector graphics will soon be readily available as SVG (see Section 4.4.5).
❑ Java applets can be used to implement many interactive features, including:

- context-dependent menus;
- advanced geographic selection (e.g. rubber-banding).
- Java has a full set of image processing classes enabling efficient operations to be carried out client side.

❑ Helper applications allow many processes to be off-loaded to other programs. This even includes some GIS operations. Other examples include:

- Downloading tables as spreadsheets;
- Passing images to suitable viewing programs.

Most of the above features can be employed to help implement an online GIS. In this chapter, we explore some of the ways in which this can be done. For the most part we focus on how to use standard Web tools and facilities. However, it is important to bear in mind that commercial GIS packages are likely to provide many tools that will greatly expand (and simplify) the range of what can be achieved.

4.2 IMAGE MAPS

One of the first systems for implementing geographic queries was the imagemap. An imagemap is an interactive image. In a simple hypertext link, clicking on an image would request a single document. However, the imagemap calls up different documents, depending on where on the image you click.

A very useful way of retrieving information is to plot data on a diagram or map that shows how ideas, places etc. are related to one another. The imagemap construct allows us to use such diagrams as hypertext indexes. The source image may be a map, but can also be any image at all, such as a diagram.

For example, the following image (Fig. 4.1) might serve as a simple geographic index to Australian states and cities. For simplicity, we have implemented the index as a set of overlapping rectangles. However, detailed polygons representing state borders could be used if precision is important.

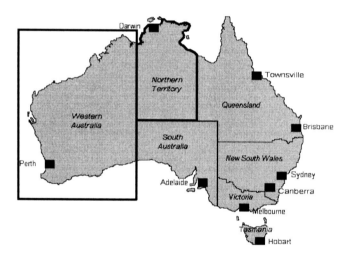

Figure 4.1. The image contained in the file `ausmap.gif`. The box around Western Australia and polygon bordering the Northern Territory indicate the regions used in the imagemap example discussed in this section.

4.2.1 Operation of an imagemap

An *imagemap* requires three elements.

1. An image (e.g. "`file.html`") so that the map can be displayed.
2. An HTML file containing the image and linking the image with the map. To be active the image must be referenced via the ISMAP syntax.
3. A map table which defines regions on the image and lists what action to take for each.

In the above example (Fig. 4.1), the image is a map of Australia (as a raster image in GIF format). The associated map file (aus.map) is shown below.

```
default /links/ozerror.html
rect /links/wa.html 0,90 219,300
poly /links/nt.html
218,114 218,248 327,248 328,133 324,128
315,123 311,124 304,116 299,115 295,110
304,101 302,96  303,91  310,89  315,78
308,75  294,79  287,75  276,75  267,69
264,69  261,78  240,79  231,88  231,94
227,94  223,105 225,110 225,114 218,112
```

This first line defines the default action – what to return if no valid region is selected. The remaining lines define regions of the map and the action to take. For example in the first line:

`"rect"`	means that the region is a rectangle;
`"./links/wa.html"`	defines the file to retrieve when the region is selected; and
`"0,90 217,370"`	defines corners of a rectangle within the image.

Note that this example has been simplified to avoid printing pages of code! For some applications, the rectangle shape, as shown for Western Australia, would be suitable, but if precision is required, then the rectangles would be replaced by polygons, as shown here for the Northern Territory, that track the state borders more accurately.

The syntax for calling an imagemap differs from ordinary hypertext. For example to use the above image as a simple hypertext link, we would use the syntax

```
<IMG SRC="ausmap.gif">
<A HREF="/links/ozweb.html">
```

However, to define the imagemap we use the following ISMAP syntax. Below, the attribute "/cgi-bin/imagemap/ausmap" denotes the map file and "ausmap.gif" is the image.

```
<A HREF="/cgi-bin/imagemap/ausmap">
<IMG SRC="ausmap.gif" ISMAP></A>
```

For more detailed instructions see an online imagemap tutorial (e.g. NCSA 1995).

4.2.2 Client-side image maps

Originally imagemaps were available only for server-side processing. That is, for every selection the coordinates were transmitted back to the server, which then performed the lookup of the map table. However, this procedure meant that imagemaps could not be used for stand-alone reference.

The browser program Netscape version 2 (1995) and later introduced client-side imagemaps, in which the browser itself processes selections on the image (there are many advantages to this approach).

The following map (Fig. 4.2) is a client-side image map. If you click on Australia, then a small map of Australia is loaded; if you click elsewhere a default file is loaded.

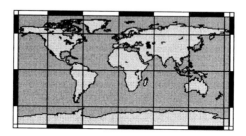

Figure 4.2. The world imagemap described in the text.

The above innovation requires two additions to HTML syntax. First the parameter USEMAP tells the browser to look for the map table at the location indicated (usually within the same document) and the <map> element defines a block of code where the table is located. Here is the code used in the above example:

```
<img src="world.gif" usemap="#xxx" ISMAP>
<map name="xxx">
<area coords="232,87,266,117" href="aus.htm">
<area coords="0,0,300,154"    href="error.htm">
</map>
```

In the above fragment of HTML code, the first line loads the image and activates the map. The remaining lines (which need not follow immediately, even though they do here) define the map table.

4.2.3 Production of imagemaps

As we saw above, define hot regions on an image by listing the coordinates of the region's border. However, these are image coordinates, not geographic coordinates. So when creating an imagemap, it is necessary to convert borders so that they refer to the image, not the geography. There are two distinct ways of doing this. If the geographic coordinates are known, then they need to be converted using the appropriate transformation. However, if they are not, then the borders need to be digitised directly. Most image viewers allow the user to read coordinates directly off an image. However, several share ware programs allow the user to build a map file directly.

4.3 USE OF JAVASCRIPT IN CLIENT SIDE OPERATIONS

Javascript is a scripting language designed for use with Web documents. It began as a proprietary Netscape invention, but has now become a web standard as ECMA script. The code can be included within an HTML document to perform various functions. The main kinds of applications are:

- ❑ To improve the user interface.
- ❑ To validate form data prior to submission.
- ❑ For animations and other "bells and whistles" (for example, animations).
- ❑ To allow interaction and exploration of content (e.g. simulations and games).

Note that Java and Javascript are quite different languages, although they share some syntactic sugar. A simplified way to view the difference is to think of Javascript as knowing what is on the Web page and being able to manipulate it. Java, on the other hand, runs in an independent window or within the web page but does not normally access its fields.

In the following sections we examine two useful examples.

4.3.1 Screening data input fields

One of the greatest problems with any large scale data system is to ensure that all of the entries are valid and in a standard format. The best place to trap errors is at data entry. Web browsers provide two ways of reducing input errors:

- ❑ Controlling values in form fields; and
- ❑ Using Javascript functions to screen data input fields.

The first of the above methods is simplest. We ensure that values are entered in the correct format by controlling what is entered. To understand this, suppose that we set up a simple form in which every value is entered as a simple text field. The problem is that people can enter values in many different ways. For instance, the name "United States of America" can be written in many ways, such as:

- ❑ United States
- ❑ USA
- ❑ U.S.A.
- ❑ US of A

Whereas humans can readily identify all of the above forms as variations of the same thing, they pose severe problems for computer searches, indexing and many other automated functions. To reduce this problem we can use use pull menus, radio buttons and other similar data fields that require the user to choose the value, rather than enter it. By adopting this method, we can ensure that the system enters the value in correct. For instance, in the example above, USA can be rendered as an abbreviation. Every time it is selected the machine enters it in the same way. Figure 4.3 shows a simple form of this kind. The HTML code that generates the form is listed below. Notice that although country names are written

Figure 4.3. WWW form resulting from the HTML code listed in the text. Although country names are written on screen in full, the values returned for the variable "country" are all abbreviations.

on screen in full, the values returned for the variable "country" are all abbreviations. This can be seen in the OPTIONS fields listed in source code.

```
<html><body>
<h1>Choose a country</h1>
<form action="http://life.csu.edu.au/cgi-bin/gis/opt1"
      method="POST">
<select name="country">
      <option value="USA">United States of
               America
      <option value="AUS">Australia
      <option value="UK" >United Kingdom
</select>
<p><input type="submit"><input type="reset">
</form>
</body></html>
```

It is advisable to adopt the above approach as widely as possible in data entry forms. For instance, to enter dates we can provide a list of all days of the month, as well as a list of months of the year. In general, text entry can usually be avoided wherever there is a finite (and relatively small) number of possible values for a given field.

Another way to trap errors using selection lists is to provide redundant fields. For instance, suppose that the user is required to enter the name and lat/long of a town in text fields within a data entry form. Then it is useful to include a field for correlated geographic units, such as country, state or province as well. This makes it possible to carry out redundancy checks and flag possible errors. For instance, the town name could be checked against a gazetteer of place names within the selected geographic unit. Likewise the latitude and longitude can be tested to confirm that they do lie within the selected unit.

4.3.2 The need for Javascript

The above method of providing choices during data entry applies only to fields that have a restricted range of possible values. However, it cannot trap all problems with data input. For example, the user may fail to *select* a value where one is required. And inevitably free text will be required in almost any data entry form. For text fields we can use Javascript to carry out preliminary screening of input data before it is transmitted to a server.

Figures 4.4 and 4.5 show an *example* of a simple form for data entry. In this example the user enters the name of a town and the latitude and longitude of its location. The code for the form includes the required Javascript, which is downloaded to the browser, along with the form. The field

```
ONSUBMIT="return checkForm()"
```

within the FORM tag instructs the browser to run the Javascript code just before submitting it to the server. The form data is transmitted only if the function checkForm returns a value of TRUE, otherwise an alert box appears with details of the error detected (Fig. 4.5). The following code shows how the Javascript is integrated into the HTML code that produces the form.

```
<HTML>
<HEAD>
<SCRIPT LANGUAGE="JAVASCRIPT">
//--Check the submitted form
function checkForm()
{
    if ( (isLatLong()) && (isPlaceName()) )
    {
      return true;
```

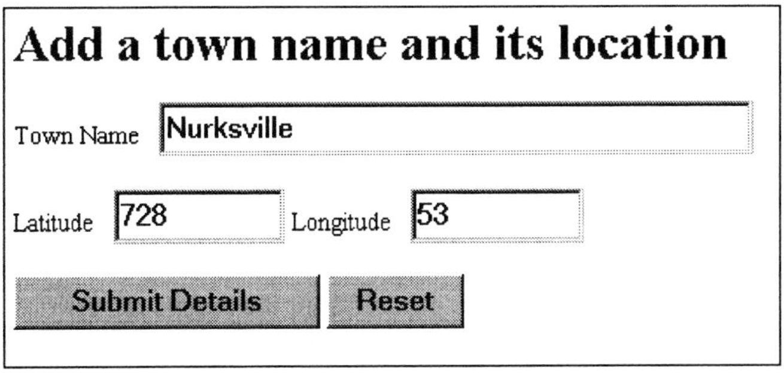

Add a town name and its location

Town Name Nurksville

Latitude 728   Longitude 53

Submit Details   Reset

Figure 4.4. A simple form for entering town locations. The source code contains Javascript (partly omitted) to test the validity of the entries prior to submission. The missing code is shown in the next figure.

```
      }
      else
      {
        return false;
      }
}
        ... code omitted for isLatLong and isPlaceName ...
</SCRIPT>
</head>
<body>
<FORM ACTION="form_processing_url"
      METHOD="POST"   NAME="data_entry"
      ONSUBMIT="return checkForm()">
<h1>Add a town name and its location</h1>
<INPUT TYPE=HIDDEN NAME="topic" VALUE="">
<p>Town Name <INPUT TYPE="text" NAME="town" SIZE=40>
<p>Latitude  <INPUT TYPE="text" NAME="latitude"
SIZE="10">
<p>Longitude <INPUT TYPE="text" NAME="longitude"
SIZE="10">

<INPUT TYPE="Submit" VALUE=" Submit Details ">
<INPUT TYPE="Reset" VALUE=" Reset ">

</FORM>
</body></html>
```

In this example, we have applied three different kinds of checks.

❑   The first is to check that the name field is not empty. This is a useful way to ensure that required data fields are completed prior to submission. We can generalise the code used as follows, where angle brackets indicate variables that should be replaced with appropriate field names or code.

Figure 4.5. The javascript source code missing from the form in the previous figure. This code checks that the fields are valid. An error produces an alert, as shown.

```
            if (<field_name> == "") { <action> }
```

❏ The second check ensures that the town name is of an acceptable form. For simplicity, the code used in the example given here

```
    var ch = townstr.substring(i, i + 1);
    if ((ch < "a" || "z" < ch) && (ch != " "))
```

is very restrictive. It ensures that each character in the name is either a letter, or a space. In real applications, a much greater variety of characters would be permissible.

❏ The final check is to ensure that values are given for latitude and longitude and that they lie within the permissible range. In this case we have used the convention that positive valus for latitude denote degrees north and that negative values indicate degrees south.

```
    if (townlat =="" || townlat < -90 || townlat > 90 )
```

The source code for carrying out the above tests is included in the Javascript code that follows.

```
// Checks town field.
function isPlaceName ()
{
var townstr = document.data_entry.town.value.toLowerCase();
// Return false if field is blank.
    if (townstr == "")
    {
        alert("\nThe Town Name field is blank.")
        document.data_entry.town.select();
        document.data_entry.town.focus();
        return false;
    }
// Return false if characters are not letters or spaces.
    for (var i = 0; i < str.length; i++)
    {
        var ch = townstr.substring(i, i + 1);
        for (var i = 0; i < townstr.length; i++)
        {
            var ch = townstr.substring(i, i + 1);
            if ((ch < "a" || "z" < ch) && (ch != " "))
            {
            alert("\nTown names may contain only letters
                    and spaces.");
            document.data_entry.town.select();
            document.data_entry.town.focus();
            return false;
            }
    }       return true;
}
```

```
//Check Lat & Long fields
function isLatLong ()
{
     var townlat   = document.data_entry.latitude.value;
     var townlong = document.data_entry.longitude.value;
// Return false if latitude  is outside the range -90,90,
// or              if longitude is outside the range -180,180.
     if (townlat =="" || townlat < -90 || townlat > 90 )
     {
       alert("\n Latitude must be in the range -90 to 90");
       return false;
     }
     if (townlong =="" || townlong < -180 || townlong > 180)
     {
      alert("\n Longitude must be in the range -180 to 180");
      return false;
     }
     return true;
}
```

4.3.3 Geographic indexes

Javascript can also be used for many other purposes. For instance we can enhance image maps by eliminating the need to refer back to the server. In the example shown (Fig. 4.6), we replace calls to the *server* with calls to a Javascript function. Whenever a mouse click is made on the imagemap, this function generates a fresh document and inserts the appropriate text into it. The following Javascript code provides the functionality for this example.

```
<HTML>
<HEAD>
<SCRIPT LANGUAGE="JAVASCRIPT">
//--Return selected information
function DataDisplay(map_option)
{
    msgWindow=window.open("","displayWindow",
       "menubar=no,scrollbars=no,status=no,
       width=200,height=100")
    msgWindow.document.write(
       "<HEAD><TITLE>Data display<\/TITLE><\/HEAD>")
       if (map_option==1)
       {       msgWindow.document.write(<h1>Australia<\/h1>')
               msgWindow.document.write('Data about
Australia<br>')
       }
       if (map_option==99)
       {
               msgWindow.document.write('<h1>Try
again!<\/h1>')
```

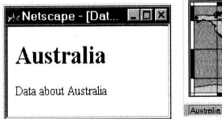 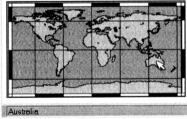

Figure 4.6. Combining an imagemap with javascript to produce a simple spatial information system. In this simple example a data window appears (shown at right) when the user clicks on the relevant area of the map.

```
        }
}
</SCRIPT>
</head>
<body>
<h1>Information Map</h1>
<img src="world.gif" usemap="#xxx" ISMAP>
<map name="xxx">
<area coords="232,87,266,117"
        href="Javascript:DataDisplay(1)"
        onmouseover="self.status='Australia';return true"
        onmouseout="self.status='';return true"
>
<area coords="0,0,300,154"
        href="Javascript:DataDisplay(99)"
        onmouseover="self.status='Try again';return true"
        onmouseout="self.status='';return true">
</map>
<p>
</body></html>
```

An important implication of the above is that a number of indexing and other operations can readily be transferred from the server to the client. This is desirable where practical, unless it would compromise proprietary or *other* concerns. Given that Web sites often contain images that total several hundred kilobytes in size, it is not impractical to download data tables of hundreds, and perhaps even thousands of entries.

Javascript can be used to make complex mapping queries self-contained. For example, Environment Australia's online "Australian Atlas" (Fig. 4.7) uses Javascript to allow users to select layers and other options in the course of making customised environmental maps.

4.3.4 Other applications of Javascript

Javascript provides a means to transfer a very wide range of operations from server to browser. For example, continuity in an interactive session means that the form or dialog to be displayed depends on the *choices* and inputs made by the user. This process is slow if the documents must be repeatedly downloaded from the server.

4.3.5 Cookies

A "cookie" is an HTTP header that passes data between a server and a client. The main function of cookies is to help a *server* to maintain continuity in its interactions with users. As we have seen, HTTP is memoryless. Every interaction is independent of previous interactions, even if a user is browsing through a site during a single interactive session. One of the motivations for cookies was to be able to facilitate interactive sessions by providing a means for a server to keep track of a user's previous activity.

The core of a cookie is the name of a variable and its value. Usually this variable provides a user identification, which allows the server to look up relevant background details on the history of the user's previous *interactions*. However, the value could be anything at all. For instance, in the context of a GIS session it could

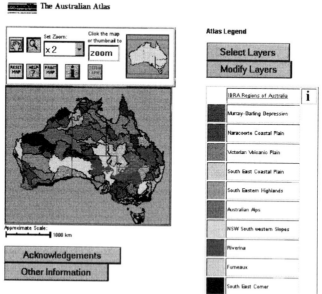

Figure 4.7. A server-side form interface to Environment Australia's Australian Atlas. Javascript functions allow the user to select and change layers and features to be displayed.

encode details of the type of operation that the user is currently undertaking, although this information is often more simply passed as hidden form fields. The typical format for cookie *header* is as follows:

```
Content-type: text/html
Set-Cookie: user=Nurk01; path=/gis;
            expires Sat, 13-Apr-2002 12:15:00 GMT
```

Table 4.1 sets out the meaning of the fields used in this header, as well as other fields assumed in the process. To *maintain* continuity from one session to another, cookies are usually written to a cookie file that is stored on the hard disk of the machine running the Web browser concerned.

Table 4.1. Examples of entries from the file cookies.txt

Field	Meaning
Name=value	The variable name and its value (here user=Nurk01)
Path	The path to which the cookie applies (here it is /gis)
Expires	When the cookie ceases to be valid (expressed here as Greenwhich date and time).

Table 4.2 shows two simple *examples* of entries stored in a user's cookie file, which would be referenced by a typical web browser. In practice the entries would be laid out in TAB delimited format in a flat text file.

Table 4.2. Examples of entries from the file cookies.txt

Domain	Flag	Path	Secure	Expiry	Variable	Value
.gis.org.au	FALSE	/gis	FALSE	1578832501	user	Nurk01
www.cookiecentral.com	FALSE	/	FALSE	978407300	foo	bar

The domain field specifies the *server*(s) to which the cookie applies. The two logical fields are set by the user. The FLAG field defines whether all machines in the given domain can access it. The SECURE field defines whether the server needs to provide a secure connection before it can access the entry.

Cookies can be created and retrieved using Javascript (as well as other commonly used languages such as Perl and *VBScript*). In Javascript there is a default object **document.cookie**, which handles interactions with cookies. Passing a cookie to this object causes it to be created and stored.

The following function (Whalen 1999) retrieves a cookie from the object document.cookie.

```
function getCookie(name) {
        var cookie = " " + document.cookie;
        var search = " " + name + "=";
        var setStr = null;
        var offset = 0;
        var end = 0;
        if (cookie.length > 0) {
```

```
                        offset = cookie.indexOf(search);
                        if (offset != -1) {
                                offset += search.length;
                                end = cookie.indexOf(";", offset)
                                if (end == -1) {
                                        end = cookie.length;
                                }
                                setStr =
        unescape(cookie.substring(offset, end));
                        }
                }
                return(setStr);
        }
```

To use this function the user would call the *function* with the name of the relevant cookie variable. For instance to retrieve the cookie "user" from Table 4.2, a possible call is:

```
        gis_user_var = getCookie( "user");
```

4.4 THE USE OF JAVA APPLETS

Java is a fully fledged programming language for the Web. It was developed by SUN Microsystems to meet the need for a secure way of introducing processing elements into HTML pages. It is object-oriented and includes a wide variety of graphics and other functions.

Like Javascript, the Java language can be used to provide client side functionality. Many of the functions described above for *Javascript* (e.g. context-sensitive menus) can also be implemented using java. However, another important use for java is its drawing ability. Here we look at two examples of interactive operations that are fundamental in GIS, but difficult to implement in other ways: rubber banding, tracing polygons and drawing maps.

4.4.1 Drawing

An important use of Java is to reduce the processing load on a Web server by passing the task to the client. Another is to reduce the volume of data (especially large images) that need to be passed across the network. One of the heaviest processing loads (when repeated on a large scale) is simply drawing maps. When done on the server, there is both the time cost of processing and a network cost in the form of maps presented as large images that need to be transferred from server to client. In many instances it is both faster, and involves less data transer, to pass raw data to the client, together with the Java code needed to turn it into a map.

The following fragment of Java code is a simple example of a function that draws a map as an image. In this case it also draws a rectangle on the map to indicate a selected area. As we shall see later, SVG now provides an alternative.

```
public void paint( Graphics g )
{
    g.drawImage( mapImage, 0, 0, mapWidth, mapHeight, this );
    width = Math.abs( topX - bottomX );
    height = Math.abs( topY - bottomY );
    upperX = Math.min( topX, bottomX );
    upperY = Math.min( topY, bottomY );
    g.drawRect( upperX, upperY, width, height );
}
```

4.4.2 Selecting regions

Earlier we have seen two ways of making geographic selections via a standard Web browser. The first was to use an image within a form as a data entry field. This method allows the user to enter the *image* coordinates of a single point within a map. The second method was to define an imagemap within a standard HTML document. The imagemap construct allows users to select a pre-defined region from a map.

The above types of selection omit several important GIS operations.

❑ Although in principle the form method could be used to select a variety of objects (e.g. a road), in practice the user could not confirm that the correct object had been selected without reference back to the server, which is time-consuming and clumsy.

❑ Selecting arbitrary, user-defined regions requires the user to be able to draw a polygon by selecting a series of points.

❑ Single point clicks suffice for defining single points. However, to define a set of points, or to digitise (say) a line, such as a road or river, or a region, the user must again be able to select a set of points and have lines joining them drawn in.

❑ Dynamic movement is an interactive process in which a user selects an object and moves it. This is useful to denote (say) the changing position of a car on a road. It is closely related to rubber banding.

❑ Zooming and panning involve redrawing a map interactively.

In the following sections we look at *some* of the methods needed for carrying out these functions.

4.4.3 Rubber banding

Rubber banding is the process of selecting a region (usually a rectangle, circle or some other regular shape) by choosing a point and holding down the mouse button whilst simultaneously "pulling" the shape out until it reaches the desired size (Fig. 4.8).

In practice, rubber banding is an example of interactive animation. That is, the maps image, with the shape overlaid on it, is repeatedly redrawn and displayed in response to the user's mouse movements. Conceptually this procedure involves the following steps:

```
Get mouse location
Calculate shape coordinates
Copy image of base map
Draw shape image over base map
Redisplay map image
```

The following java source code illustrates some of the functions required in the above example. This code is used in conjunction with the earlier listing (Section 4.4.1) that drew a map with a rectangle overlaid on it. Notice that the listing includes functions for dealing with three distinct events involving the mouse: that is, the mouse being *pressed*, *dragged*, and *released*.

```java
public void mousePressed( MouseEvent e )
{
    xPos = e.getX();
    yPos = e.getY();
    mousePosition = myMap.checkPosition(xPos, yPos);

    if(mousePosition)
    {
        setTopX( xPos );
        setTopY( yPos );
        tempUpperLong = myMap.calcLong(xPos);
        tempUpperLat  = myMap.calcLat(yPos);
    }
    else
    {
        showStatus("Mouse is outside the map area");
    }
}
```

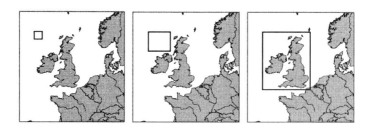

Figure 4.8. Rubber banding.

```
public void mouseReleased( MouseEvent e )
{     xPos = e.getX();
      yPos = e.getY();
      if(rodentDrag == 1 && rodentRelease == 0)
      {
          setBottomX( e.getX() );
          setBottomY( e.getY() );
          mousePosition = myMap.checkPosition(xPos, yPos);
          if(mousePosition)
          {
              lowerLat   = myMap.calcLat(yPos);
              lowerLong = myMap.calcLong(xPos);
              upperLong = tempUpperLong;
              upperLat   = tempUpperLat;
              myMap.convertToPixels(upperLong, lowerLong, u
                                    upperLat, lowerLat);
              zoom.setVisible(true);
              rodentRelease = 1;
              mousePosition = false;
          }
          else
          {
              countryInfo.setText("Outside map. Try again\n");
                            mousePosition = false;
          }
      }
}

public void mouseDragged( MouseEvent e )
{     if(mousePosition)
      {     setBottomX( e.getX() );
            setBottomY( e.getY());
            rodentDrag = 1;
            repaint();
      }
      else
      {     showStatus("Mouse is outside the map area");
      }
}
```

4.4.4 Drawing and plotting maps, graphs and diagrams

The final application we consider here is the use of java to draw entire maps on the client machine. The ability to draw polygons interactively is crucial in GIS. It is used in both drawing and selecting geographic objects. The ability to draw and interact with maps on the client machine has several potential advantages:

❑   For vector maps, it usually requires less data to download the vector data that defines a map, than to download an image of the map. It also reduces server side processing.

□   Client-side drawing cuts down the amount of processing required of the server.

□   It speeds up processes such as zooming and panning, which otherwise require the constant transfer of requests and responses between client and server.

The need to develop java applets to deal with the above issues may decrease as the World Wide Web Consortium introduces new standards. We now look at one of these in the following section.

4.4.5 Scalable Vector Graphics (SVG)

Methods of drawing and plotting maps and other online figures are likely to change dramatically with the introduction of new standards and languages for plotting vector graphics on the Web. One of the problems with online graphics has been that Web browsers could display only raster images, usually in GIF or JPEG formats. However, pixel based images suffer from the problem that they cannot be rescaled, which is a problem when printing screen images. They also tend to lose resolution from the original. Diagonal lines, for instance, often exhibit a "staircase" effect. Another huge problem is that pixel-based images can produce very large files. This is one of the biggest factors in slowing delivery of online information, especially for users linked via modems. Given that most GIS is vector based, the necessity of delivering pixel images has been a great nuisance.

In February 1999, the World Wide Web Consortium released the first draft of a new standard covering *Scalable Vector Graphics* (SVG). The following details are taken from the draft specification (Ferraiolo 2000). At the time of writing SVG is still only a candidate recommendation, and may change before being finally adopted. These examples use some of the XML markup features described more fully in Chapter 5.

To enable SVG an HTML document should include a header such as the following (Ferraiolo 2000).

```
<!DOCTYPE svg PUBLIC "-//W3C//DTD SVG 20001102//EN"
  "http://www.w3.org/TR/2000/CR-SVG-20001102/DTD/svg-
20001102.dtd">
```

The language includes syntax for describing all the common constructs and features of vector graphics. It also recognises the need for integrating graphics with hypermedia elements and resources. This ability is essential, for example, to create imagemap hot spots in an image, or to pass coordinates to a GIS query.

Because polygons defining regions are so common in GIS, perhaps the most appropriate example is the following, in which the code defines a simple filled polygon drawn inside a box, much as a simple map might be. Example **map01** specifies a closed path that is shaded grey. The following abbreviations are used for pen commands within the path command:

```
M       moveto,
L       lineto
z       closepath
```

Using these pen commands, the SVG code for example **map01** is as follows.

```
<?xml version="1.0" standalone="no"?>
<!DOCTYPE svg PUBLIC "-//W3C//DTD SVG 20001102//EN"
   "http://www.w3.org/TR/2000/CR-SVG-20001102/DTD/svg-
     20001102.dtd">
<svg width="4cm" height="4cm" viewBox="0 0 400 400">
   <title>Example map01 - a closed path with a border.</title>
   <desc>A rectangular bounding box</desc>
   <rect x="1" y="1" width="500" height="400"
         style="fill:none; stroke:black"/>
   <path
     d="M 200 100
     L 300 100
     L 450 300
     L 350 300
     L 300 200
     L 300 300
     L 100 300
     L 100 200
     L 200 200
     z"
     style="fill:grey; stroke:black; stroke-width:3"/>
</svg>
```

Although we have deliberately kept this example brief, it is easy to see how the method could be used to draw, say, the coastline of North America. Note that we can readily integrate SVG code into the object-oriented method discussed in Chapter 1. For instance, each object, say North America, would have associate with it methods to draw a graph of itself, either as a list of coordinates or as calls to

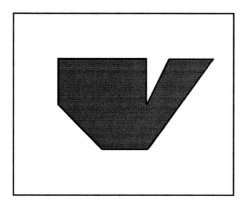

Figure 4.9. The figure produced by the SVG code given in the text.

sub-objects (e.g. Canada, USA). The method might include the following steps for each object:

- ❑ convert the list of lat-long bounding coordinates into SVG path commands.
- ❑ print the name of the object by writing an SVG text command

Creating an entire map from a description of the data objects is simply a matter of working recursively through the hierarchy of objects in the map. At each stage, you convert each object into its SVG representation. The process may sound cumbersome, but it is the kind of processing that computers thrive on.

Because SVG is based on XML (see Chapter 5), there is ample scope to bundle up different map elements as objects wrapped in the corresponding XML elements. So for instance, we could render the above example as a hierarchy of map objects, with the box indicating the border of the entire map, and the path indicating one feature in the map.

The compact nature of vector graphics means that down-loading even complex maps would still require less data than an image of the same area.

We anticipate that the introduction of SVG is likely to have several effects. One is that the need for java code as a rendering tool for maps will be reduced. SVG viewers are likely to include of the operations (e.g. zooming and panning) described earlier. So in many instances a map can be downloaded from a server as SVG source and manipulated directly.

Third party packages are likely to include scripting commands and other high level features to simplify the generation of SVG source. One important step would be converters for turning files in proprietary GIS formats into SVG code. Many SVG related tools are already available. Most basic are programs and plug-ins for viewing SVG data Other basic tools include filters for converting translating data between SVG and several GIS and CAD/CAM formats. They also include converters for SVG into standard image and output formats, such as GIF, Postscript and PDF.

4.5 EXAMPLES

4.5.1 An example of an interactive geographic query system

As a simple demonstration of the use of Java applets, we have implemented an online query system for Australian towns (http://life.csu.edu.au/gis/???). In this system the user downloads an applet that displays a map of Australia together with dots for towns (Fig. 4.10). To obtain information about a particular town, the user selects a town by mouse click (Fig. 4.10a). When a town is selected in this way, the details are displayed in boxes on the screen. The user can zoom into particular regions by rubber banding (Fig. 4.10b).

4.5.2 Asian financial crisis

The company Internetgis.com has developed a system, called ActiveMaps, which provides a class library is designed for Java developers who want to add GIS/mapping functions to their applets or applications. The service provides geographically based information about the Asian financial crisis. It combines a map of the region with data about the economy of each country. The user interface consists of a map, together with a variety of tools for customisation and queries. The mapping features includes panning and zooming, using rubber banding, plus the facility to insert data on the map.

4.5.3 NGDC Palaeogeographic data

The US National Geophysical Data Centre provides large, public repositories for many kinds of palaeoclimatic, such as tree ring records, and the Global Pollen Database (NGDC 2000). To assist users, NGDC have developed a Java Applet called WebMapper for finding Paleoclimate data held at the World Data Center for Paleoclimatology (Fig. 4.11). The service includes zooming, site selection. Enhancements planned (at the time of writing) include the ability to search for sites by name or by investigator, or to filter the sites displayed by type, age range, variable, or investigator.

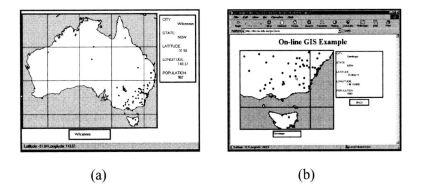

(a)   (b)

Figure 4.10. A demonstration of an online geographic query system using Java. The Australian towns query system allows users to (a) the online user interface, which is invoked when the system is started. (b) An example of zooming (using rubber banding) and displaying selected data. http://life.csu.edu.au/gis/javamap/

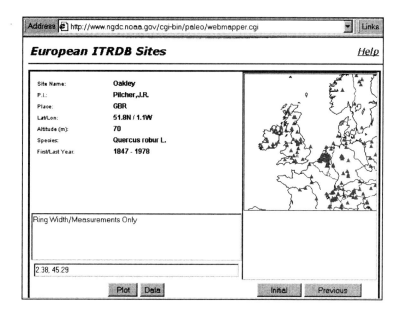

Address 🔲 http://www.ngdc.noaa.gov/cgi-bin/paleo/webmapper.cgi   ▼ | Links

European ITRDB Sites   *Help*

Site Name: **Oakley**
P.I.:   **Pilcher,J.R.**
Place: **GBR**
Lat/Lon: **51.8N / 1.1W**
Altitude (m):   **70**
Species: **Quercus robur L.**
First/Last Year: **1847 - 1978**

Ring Width/Measurements Only

2.38, 45.29

Plot   Data   Initial   Previous

Figure 4.11. The WebMapper search interface (Java Applet) from the World
Data Center for Paleoclimatology. http://www.ngdc.noaa.gov/paleo/data.html

4.6 CONCLUSION

In the last two chapters we have described the basic Web technology that can be
adapted to create an online GIS. We should perhaps stress that the techniques that
we have described here are based on free software and can be used as an integral
part of a web site without incurring the start up and maintenance costs of a
commercial GIS.

An important area of interest for minimalist client-side operations is the
palm-top market. At the time of writing, this market is set to explode. Several
commercial GIS manufacturers have recently introduced palm-top extensions. We
will discuss this development later (see Chapter 11).

In the chapters that follow, we now turn to look at ways in which individual
online GIS can be integrated into more comprehensive geographically based
information systems.

## CHAPTER 5

Introduction to Markup

5.1 MARKUP LANGUAGES

In the early days of computing, formatting data was pretty much up to the idiosyncrasies of individual programmers. If you wanted to be able to plot the boundary of a country, then you might throw the data into a file as line after line of data, each line containing a pairs of coordinates. The file would contain numbers, just numbers. As likely as not, the coordinate pairs would be jammed up against one another. Clarity didn't matter because the data were set up for just one program to read and you told your program what the exact format was. If it ever occurred to you that others might one day want to use that data, then you just shrugged your shoulders and said "Hey, that's ok. If they need to know what the format is, then just come and ask me."

Today's computing scene is completely different. Data and processing are distributed across networks. An online GIS is likely to need to draw on data from many different sites. Likewise, any data set is likely to accessed and used by many different programs. Given this environment, there is no time to go and ask the author what the format is. The content, structure and format of a data file needs to be immediately transparent to any program that accesses it. This need creates a demand for ways of indicating the structure of data. But *ad hoc* explanations are not enough. They need to be universal standards that everyone can understand and apply. And so markup languages were born.

In the very early days of computer document processing, full screen displays of a document as it would look when printed, so-called *what you see is what you get* (WYSIWYG), was far from practical. Not only were machines not powerful enough, either in processor speed or in memory, to manipulate documents on the fly, but computer monitors were hardly up to the task. Windowing systems, universal today, were still to come. Yet there was a need to go beyond plain text, just an endless list of words with simple punctuation, to text that had different fonts or structure (numbered paragraphs, chapters, indexing and so on).

Thus began the idea of embedding special characters, called *markup tags*, in the text. Markup tags introduced, not more text, but control sequences. Early examples of such systems were the utilities `nroff` and `troff` on Unix and `runoff` on DEC machines. In `runoff`, for example, a dot at the beginning of the line signified that the line was not part of the text, but contained commands to the processing system. Once a document had been written up with the associated commands, or markup, a program would read it, put the commands into action and produce a printable form.

A major breakthrough in computer typesetting occurred in the late 70s, early 80s, when Donald Knuth (1984) developed the language TeX. Designed to facilitate the typesetting of mathematics (and improve the accuracy in the process), TeX introduced a whole range of new concepts and methods for representing and processing typographical data. TeX is still a markup language with a very, very powerful macro facility (i.e. a system for storing and re-using sequences of commands); but it is concerned primarily with layout and presentation. Moving a full stop a fraction of a millimetre is a lot of fun in TeX!

TeX is, however, complex! Users frequently relied on collections of macros or *ad hoc* templates from elsewhere. The successor language LaTeX, which followed in around 1986 (Lamport 1986), introduced an integrated macro framework using pre-written style files. It was much easier for the non-specialist to use, but its really important contribution was its emphasis on structure. A LaTeX document consists of a range of nested environments, nested within one another. From the document environment downwards they indicate the structure of the document as sections, lists, paragraphs etc. For research papers in the quantitative sciences, there is still no significantly better typesetting system.

Although nroff and troff are still often found on Unix machines, most of these systems are now obsolete. But one, the Generalised Markup Language from IBM, evolved into SGML, the *Standard Generalised Markup Language*. which subsequently gave birth to XML. Originating around the same time as TeX, the emphasis in SGML was, right from the beginning, on structure rather than presentation.

SGML was a well-kept secret for many years. In fact one book was even called the *Billion Dollar Secret* (Ensign 1997). It had some big customers: you don't get much larger than the US military. SGML, the Standard Generalised Markup Language is not in itself a markup language, but rather a *meta-language*: it enables precise document specifications, or DTDs (Document Type Definitions) to be written. The merit of such a scheme is that it becomes possible to precisely control the structure of a document. The military found this particularly desirable having vast quantities of paperwork specifications and a need for powerful document control techniques.

SGML really became a household tool with the beginning of the World Wide Web, although, with apologies to Voltaire, using SGML was like writing prose for many people. What everybody saw, of course, was not SGML but one DTD written within it: HTML, the *Hypertext Markup Language*. Early exponents of the Web followed HTML quite closely, but the Web grew at a pace hardly anybody dreamed of. HTML 1.0 was followed rapidly by ever more complex versions. It is currently at level 4, which is at least four times the size of the original HTML specification!

The World Wide Web now covers almost everything imaginable: from academic papers to advertising; from pornography to space science; from the trivial and ephemeral to ground-breaking, new research. In 1999 the number of Web sites hit somewhere around 50 million, with over 200 million forecast for the year 2000. With this explosion has come a truly enormous problem: finding things. Because the Web grows organically, without any central control, it is inherently disordered. Hence the need for *metadata*; and along with this need came the latest markup innovation, XML.

Originally SGML was concerned primarily with document structure. SGML would ensure that all the pieces of the document were present and in a correct order. It's easy to see how this might be important. So, imagine competing tender documents for a new aircraft. Important items such as delivery date, price structure, and lots of other details need to be present. Why make a (fallible) human proofreader check that everything is there if it can be done by the document structuring package?

But along with the disordered growth of the Web another aspect started to assume more importance: the use of markup tags to convey *semantic* information as well as structure. Let's see how this works. Imagine that we want to find a recipe for the old English dish of jugged hare. If we look for hare we find zoological data; we find children's stories about mad March hares; we find polemics against hare coursing, one of England's less animal-welfare oriented pursuits; we might even find references to hares as food somewhere too. Now imagine that we have tagged the occurrence of jugged hare with the label dish,

```
<dish>jugged hare</dish>
```

The word "dish" in angled brackets is the tag; we'll cover the syntax in more detail below. Now if we restrict our search to hare *inside the dish tag* we get rid of most of the spurious hits. But we can do even better. We still might hit the biography of the Duke of Upper Ernest, whose favourite dish was roast hare. Suppose now that all recipes for hare are embedded within an additional tag like this

```
<recipe>
        <dish>juggedhare</dish>
        …
</recipe>
```

Now we can search for hare inside food inside recipe and we should hit just what we want. When we study the *Geographic Markup Language* (GML), which is an extension of XML, we shall see how easy it is to incorporate geographic constraints, such as recipes from a particular region. The beauty of all this, the beauty of XML, is that we don't need anything special here. We don't need an elaborately configured database, any particular metadata standard will do.

But there could be one complication. Different people might use tags in different ways. Dish could be used to tag pottery. Thus we have come to the idea of *namespaces,* which allow one to register the use of a tag. We'll cover this in Section 5.4. SGML is pretty arcane in these days of desktop publishing. XML removes a lot of the complexity (and a bit of the rigour too). But it has a free-form nature which makes it much easier to use. It was the big growth event of the years 1999–2000 and is still growing.

5.1.1 Spatial tags

We are not concerned here with all the many aspects of Web publishing. We need just enough knowledge to work through the spatial metadata standards, which are strongly dependent on SGML/XML. Now, although SGML came first and is in some respects the more powerful of the two, we will start with the more straightforward XML. XML does have one advantage over SGML for spatial metadata: it allows much more tightly controlled data types. Thus we can specify that a number shall be between 0 and 180 (such as we would require for a spatial coordinate). First we pick our way through the syntax, watching the functionality and application rather than the fine syntactic detail. Then we look at two important related standards. We consider the important issue of attaching unique interpretations or meaning to XML tags through the namespace recommendation in Section 5.4. There is now a firm recommendation for XML namespaces. The other topic is still somewhat fluid: this is the area of XML schema, which we consider in Section 5.5.

5.2 XML: STRUCTURAL IDEAS

SGML is one of the most flexible and general standards imaginable. It is composed of markup concepts and principles, but it does not specify exactly how these get represented in the source document. It does, however, have an example (commonly used) set of syntactic elements, known as the *reference concrete syntax*. We shall skip over all these details and work with the syntactic specification chosen for XML.

Conceptually, an XML document is made up of a sequence of one or more *elements*, which may be nested inside one another. The resulting document structure we refer to as a *document tree*. Each element has a single parent element right on up to the top or outermost element which we refer to as the *document root* . Elements may be qualified by one or more *attributes*. Marked up documents can get quite complicated and so there is a sophisticated abbreviation mechanism, through the use of *entities*.

Perceptive readers may already have noticed a similarity between the idea of elements in an XML tree, and the object hierarchies described in Chapter 1. XML is highly suited to representing information objects and conversely, XML elements can rightly be thought of as information objects.

Now let's examine these concepts and their associated syntax in a little bit more detail.

5.2.1 Elements and attributes

An element begins with its name in angled brackets, as in `<dish>` above. It ends with the same structure but including a forward slash immediately after the first bracket as in `</dish>`. The content of the element goes in between and can include other elements to an arbitrary extent. The content can be as simple as a word or phrase, as in our dish example. But it can be as complicated as an entire document, maybe the recipe example above, or even an entire cookery book. Sometimes an element may have no content (for example an element indicating the insertion of an object such as a graphic). In this case the closing tag can be rolled into the start tag as in

```
<image src="mypic.jpeg"/>
```

In this example, the trailing slash "/" acts as a closing tag.

An element may have *attributes*, which qualify the element in some way. So recipe might have attributes of foodtype and usage. It might look like this:

```
<recipefoodtype="game"usage="maincourse">
```

The attributes, which are essentially keyword-value pairs, go inside the start tag. SGML is one of the most flexible and general standards imaginable. It is composed of markup concepts and principles, but it does not specify exactly how these get represented in the source document. It does, however, have an example (commonly used) set of syntactic elements, known as the reference concrete syntax, such as the angle brackets above. We shall skip over all these details and work with the syntactic specification chosen for XML.

Now, in SGML, elements (along with everything else) are defined beforehand in the DTD. The DTD specifies the order that elements go in, what elements may be nested inside others, the sorts of characters elements can contain and a variety of special features which relate to how an element is processed. In XML we do not have a DTD as such. But simple rules still have to be obeyed. The most important of these is that elements should not be interleaved. In other words each element must finish entirely within the parent element, as shown in Table 5.1.

Table 5.1. Correct nesting of elements.

Correct	Wrong
`<recipe>`	`<recipe>`
`<dish>`	`<dish>`
`  recipe text`	`  recipe text`
`</dish>`	`</recipe>`
`</recipe>`	`</dish>`

5.2.2 What's in a name?

SGML defines what each name can contain quite precisely. Sometimes the reference concrete syntax is quite restrictive, e.g. names are restricted to 8 characters. XML is more fluid than SGML and commonsense will normally suffice. Any mixture of letters and characters with limited punctuation is acceptable. Why should we limit punctuation? Well, we do not want it to interfere with the control syntax. So, since angle brackets mark the beginning and end of tags, using them in a name presents complications! We would need a mechanism to quote them, to say that they were to be interpreted literally and not as indicating the start or end of a tag. We do this with entities, which we consider next.

5.2.3 Entities

In writing documents in standard ASCII text, we occasionally come across characters we don't have a representation for, particularly in non-English words. They might be accents over letters, or different letters entirely. Another situation which causes difficulty is that of the special punctuation character, such as the non-breaking space. We represent these cases using character entities. The format is simple. The entity begins with the ampersand symbol, is followed by a character string and terminated by a semi-colon. So, for the non-breaking space the character entity is . It is not surprising to find that < and > are the entities for the angled brackets while &amps; is the ampersand itself.

5.3 THE DOCUMENT TYPE DEFINITION

As we have already indicated, the DTD provides a very tight definition of document structure. We do not have the time and space to go into the fine details here, particularly since the future, especially for online documents, will be XML.

In addition to the definitions of the elements and attributes we have just discussed, the DTD contains also a certain amount of front matter. First, it contains information about the version of SGML and the locations of any files of public entities, such as special character sets and son on. Then, quite different to XML it contains definition of the syntax used for the document. SGML is truly self-describing.

Various organisations including ANZLIC and the FGDC have created DTDs for spatial metadata. We discuss them in some detail in Chapter 9.

5.4 XML NAMESPACES

As we saw earlier, XML namespaces are devices that allow us to attach meaning to XML tags. In SGML it is possible to refer to external entity sets or DTDs, but there is no precise network link. With XML, which was driven by the needs of the Web, indices to elements and attributes are, naturally enough, available as URLs[1].

XML namespaces have a precise grammar, but for present purposes we need to know just two things about their use:

1. the **xmlns** tag is used to define the URL where the namespace is to be found;
2. tags may be now indexed according to the namespace using the form **gis:bridge** to indicate a tag called bridge in the gis name space.

Tags without a namespace prefix inherit the namespace from the next outermost namespace tag. These facilities are important for the Resource Description Framework, which we look at in detail in Chapters 7 and 8. To guide

---

[1] Strictly speaking all these documents are now written in terms of the more general concept of URIs but the difference is unimportant here.

us through RDF we don't really need to know very much more, but it is worth looking at an example:

```
<section xmlns:gis="http://clio.mit.csu.edu.au/smdogis/xml">
        In the bottom left corner of the image is the
        <gis:bridge>Sydney Harbour Bridge </gis:bridge> and not
        far away is one of Australia's finest buildings the
        <building>Sydney Opera House </building>
</section>
```

There has been a certain amount of confusion over the fine details of the XML namespace definitions, which have arisen through the need to maintain backward compatibility with XML 1.0. The note by James Clark (1999) explains this in more detail but we do not need to probe further in the present context.

Also in progress is work on the XML schema requirements (Malhotra and Maloney, 1999), which will to some extent merge with the work on RDF schemas which we discuss in detail in Section 8.5.

XML allows a pretty unrestricted use of names for elements. This is fine when one author is handling relatively small documents. But for larger scale systems, web sites etc., we could run into name conflict. Names may get reused in different contexts, creating considerable confusion. The solution is the use of namespaces, defined as URIs online. Each tag may then be qualified by a namespace indicator to make its significance precise.

Thus the resource description framework, which we discuss in Chapter 8 has a namespace prefix denoted by **rdf**, hence the creator tag can be prefixed as:

```
<rdf:creator>Garfield</rdf:creator>
```

This format enables a fairly common word, such as creator, to be given a precise reference. This nomenclature could get fairly complex, but there are rules, which we do not have space to discuss here, that allow developers to set up assumptions and hierarchies of namespace references.

The SGML DTD provides a precise description of document structure: which elements go where; what element is nested within what. But a DTD says nothing about the meaning or representation of the terms. Let's eavesdrop on a conversation:

> *If the palooka sitting East had two bullets he would have doubled. So, he must have a stiff in one of the minors, we can throw him in to suicide squeeze West.*

If you've read the right books, then you might recognise some of this. The rest is jargon or slang. You'd need a glossary or thesaurus to make sense of it.

XML recognises that the various meanings of terms can change in different contexts and sets out to define them, through the concept of namespaces. A document may use more than one, externally defined namespace (somewhere on the web) and there are various defaults defined for names which are not prefixed with a namespace. We won't go into the exact details of the specification, but just look at a couple of simple examples to get an idea of what happens.

First imagine we have a namespace which defines various sailing terms, which we might label as sail. So the tag, crew, which could refer to a type of haircut, is appropriately labelled:

```
<sail:crew>Fred Bloggs</sail:crew>
```

This tells us that the term *crew* is an element in the *sail namespace*.

So how do we locate the definitions of sailing terms? We use the attribute, defined in XML, *xmlns* to provide a URI (Uniform Resource Indicator) for where the information is found.

There are two other ideas we need to grasp. The first is how we nest elements and infer default namespaces.

The second is just to recognise that we are using at least one namespace without referring to it: the XML namespace itself. One of the beauties of XML is it is incredibly self-referential. Most concepts are defined within XML as we shall see in the remainder of this and the following chapter.

Although XML design does not espouse terseness, continual addition of prefixes would start to make a document hard to read. Hence we use the concept of inheritance. A tag without a prefix inherits its prefix from the parent element, or grandparent element and so on up the document tree. Thus in the following fragment:

```
<sail xmlns="http://sailing.vir/terms">
        <dinghy>laser</dinghy>
        <cat>Stingray</cat>
</sail>
```

is equivalent to

```
<sail xmlns="http://sailing.vir/terms">
     <sail:dinghy>laser</sail:dinghy>
     <sail:cat>Stingray</sail:cat>
</sail>
```

Note that dinghy is pretty specific, but cat is not. In fact a web search for cat would have many hits which were absolutely nothing to do with boats, hence the importance of the namespace. There is an alternative syntax here, for no obvious reason, where we spell out the attribute more explicitly

```
<sail:boat xmlns:sail='http://sailing.vir/terms'>
     <dinghy>laser</dinghy>
     <cat>Stingray</cat>
</sail:boat>
```

5.5 XML SCHEMA

In SGML the Document Type Definition served to exactly encode the rules of document structure. In XML we have much looser rules, which merely control things like the embedding of tags inside one another. The need for stronger control is satisfied by XML schemas. They describe the tags and their possible values in considerable detail. As we hinted above, control over the element context may include things like the range of possible numerical values which might be taken. The schema specification is quite new and implementation of schema processing software tools is just beginning.

XML began life as a simplified version of XML. It has now come to dominate current development on the Web. With its spread into areas such as metadata, digital signatures, generic document structures and so on, the need for an additional structuring mechanism became necessary. This mechanism is the XML Schema. In some ways it is a bit of a reinvention of the wheel, in that a lot of what it does is similar to an SGML DTD. But in other ways it has gone beyond the DTD framework to allow more precise specification of structure and content.

There are many bells and whistles to the schema documents. We cannot possibly include all of them here, but the web site has the URLs for tutorial documents and the full reference specifications. What we propose to do here is to look at a metadata DTD (in fact a subset of the ANZLIC DTD) and see how the XML Schema document would express the same concepts.

We shall actually start in the middle, rather than at the beginning, to focus on the definitions of elements and their attributes. Right at the root of the document tree is the definition of the anzmeta element.

This element does not contain any actual text but merely other elements. The DTD entry to define such an element is:

```
<!ELEMENT anzmeta - - (citeinfo, descript, timeperd,
        distinfo?, cntinfo+)>
```

The two hyphens indicate whether it is possible to omit the start or end tags (in this case no; possible omission would be indicated by a letter o instead of the hyphen). In XML this situation does not arise as start and end tags are obligatory. In brackets we then have a list of elements which make up the anzmeta element. The comma separating them has a specific meaning here: the elements must appear in this order only; the ampersand symbol would be used to indicate that several elements are required but may occur in any order. Two other symbols appear: ? indicates that an element is optional; + indicates that the element must occur one or more times.

In the XML schema this looks a lot more complicated. The definition of the element is fairly simple:

```
<xsd:complexType name="anzmeta" type="anzmetaType">
```

where xsd denotes the schema namespace and the element is empty (denoted by the closing symbol />). What makes it more complicated is the type attribute. Any element may have either a simple or complex type; to users of programming languages like C or C++ this is very similar to the difference between a string or integer and a class or structure. We shall see other programming analogies later. anzmetaType is complex because it is made up of other simple or complex types.

Incidentally the name of the type could be anything. We've followed a practice common in object-oriented programming of appending a descriptor (Type) to the name of the instance. This is merely a convention. So here is the definition of anzmetaType:

```
<xsd:complexType name="anzmetaType">
    <xsd:element name="citeinfo" type="citeinfoType"/>
    <xsd:element name="descript" type="descript"/>
    <xsd:element name="timeperd" type="timeperd"/>
    <xsd:element name="distinfo" type="distinfo"
        minOccurs=0/>
    <xsd:element name="cntinfo" type="cntinfo"
        minOccurs=1
        maxOccurs="unbounded"/>
</xsd:complexType>
```

So we move recursively down through the definition of each element and its type. Note that for distinfo and cntinfo we have specified an attribute which indicates the number of times the element may occur, equivalent to the SGML? and + operators respectively. We haven't specified them for the other elements, because the defaults are adequate. The default for minOccurs is 1 and the default for maxOccurs is minOccurs, i.e. the element occurs once and only once. In the cntinfo case we have specified that the maximum number of occurrences is unbounded. We could also specify a finite number (say we going to allow up to one such element for each state or territory which in Australia would make maxOccurs 8). This is added flexibility over SGML which can not express such a precise range.

All the sub-elements, citeinfo etc. will have their associated type definitions and we do not need to go through all of them. But there are a few more features we would like to illustrate. Here is the cntinfo type (not strictly according to the ANZLIC definition)

```
<xsd:complexType name="cntinfoType">
    <xsd:element name="cntorg" type="xsd:string"/>
    <xsd:element name="cntpos" type="xsd:string"/>
    <xsd:element name="address" type=addressType,
        minOccurs=0
        maxOccurs=1/>
    <xsd:element name="city" type="xsd:string"/>
    <xsd:element name="state" type="stateType"/>
    <xsd:element name="postcode" type="postcodeType"/>
    <xsd:element name="cntvoice", type="telNumType"
        minOccurs=0
        maxOccurs=1/>
</xsd:complexType>
```

The first point to note is that although we have defined city etc. as just plain strings (built-in simple types) we have chosen to define a special type for state. Here it is

```
<xsd:simpleType name="stateType" base="xsd:string">
        <xsd:pattern value="[A-Z]{}2"/>
</xsd:simpleType>
```

The States in the USA all have two letter upper case abbreviations. Thus we define a sub-type of string which restricts strings to precisely this form. Any other string will generate an error message.

Australia has fewer states than the USA. In this case we might want to tighten things up even more and specify *only* the allowed abbreviations. We do this with *enumeration*:

```
<xsd:simpleType name="stateType" base="xsd:string">
      <xsd:enumeration value="ACT"/>
      <xsd:enumeration value="NSW"/>
      <xsd:enumeration value="NT"/>
      <xsd:enumeration value="QLD"/>
      <xsd:enumeration value="SA"/>
      <xsd:enumeration value="TAS"/>
      <xsd:enumeration value="VIC"/>
      <xsd:enumeration value="WA"/>
</xsd:simpleType>
```

Another explicit example occurs in the definition of `jurisdic` which is a sub-element of `citeinfo`. Here the jurisdictions are spelt out explicitly:

```
<xsd:simpleType name="jurisdicType" base="xsd:string">
  <xsd:enumeration value="Australia"/>
  <xsd:enumeration value="Australian Capital Territory"/>
  <xsd:enumeration value="New South Wales"/>
  <xsd:enumeration value="New Zealand"/>
  <xsd:enumeration value="Northern Territory"/>
  <xsd:enumeration value="Queensland"/>
  <xsd:enumeration value="South Australia"/>
  <xsd:enumeration value="Tasmania"/>
  <xsd:enumeration value="Victoria"/>
  <xsd:enumeration value="Western Australia"/>
  <xsd:enumeration value="Other"/>
</xsd:simpleType>
```

The first thing we can do with a schema is to specify what elements and other things belong inside an element:

```
<xsd:element name=="distinfo" type="distinfoTyp">
      <xsd:complextype name="distinfoType"
      <xsd:complextype/>
</xsd:element>
```

In the case of coordinates we might need to impose limits on the acceptable values. Latitude, for example, ranges from 0 to 90 degrees. So what we need to do is to take a simple type, integer and restrict its application.

```
<xsd:simpleType name="latitude" base="xsd:integer" >
     <xsd:minInclusive="0">
     <xsd:maxInclusive="90">
</xsd:simpleType>
```

Sometimes we might want to mix element content with some basic text data. This might work as follows:

```
<distinfo>
     The following options are available:
     <ol>
          <li> mapiinfo</li>
          <li> arcinfo</li>
          <li> OpenGiS</li>
     </ol>
     These data are also available in a variety
     of other non-standard formats.
</distinfo>
```

The schema that would represent this is:

```
<xsd:element name="distinfo"   content="mixed">
     <xsd:element name="ol">
</xsd:element>
```

Note that we cannot just mix text and elements at random. We still have to specify the order in which the elements appear even though we might intermix plain text amongst them.

Suppose we want to restrict the particular strings which might be used. There is a regular expression syntax to do something just like this. The details are complex, so let's just look at a simple example.

A common problem with much legacy data is that the origin is uncertain, i.e. we do not know the start date. We could just leave this out, or add some not-known tag. An alternative would be to make the element explicitly null. In the schema this would take the form

```
<xsd:element name="begindate"
     type = "date" nullable="true">
```

and the date element itself would look like this:
```
<begindate xsi:null="true"></begindate>
```

We can of course make the element an empty element, which we do simply by including the attribute "content=empty'".

There are two things to note about this format. First, we have applied a specific namespace (an instance of an XML schema) to the null attribute. Secondly the tag is not an empty tag, but a tag with nothing in it (which has a conventional close tag).

We might want to specify that an element is made up of a collection of elements. We have several ways of doing this: **choice**, **sequence** and **all**. The element **choice** allows just one element from a selection:

```
<xsd:choice>
        <xsd:element>
        <xsd:element name= type=/>
        <xsd:group ref=junk/>
</xsd:choice>

<xsd:group name=junk>
        <xsd:sequence>
        ...
        /xsd:sequence>
</xsd:group>
```

Note that sequence is the default anyway. With the element all we have to use all of the elements but they can be in any order, e.g.

```
minOccurs and maxOccurs
```

XML schema are somewhat more precise with mixed content models.

5.6 XQL: THE XML QUERY LANGUAGE

XQL is a generic language for querying XML documents, implemented in a number of software packages. It was essentially defined in a proposal to a W3C Query Language workshop in 1998 by Joe Lapp of webMethods and David Schach of Microsoft. The W3C has a working group on XML query languages, but at the time of writing is some way off a final submission. The XQL model illustrates the sort of things we would like to be able to do, but will have to await the precise syntax.

The full standard is likely to be more complex, allowing searches to span multiple documents, but there is not even a draft recommendation at the time of writing. A related standard, which is the *Object Query Language* (OQL). Since one prominent move on the web is towards an object-oriented model of documents, it is likely that OQL will play a role in the final standard.

XQL enables selections of subsets of a document based on XML elements, along with pattern matching for the contents of elements themselves. XQL has a number of characteristics in common with SQL. Apart from the sort of operations which can be performed, it is declarative, rather than procedural. Thus XQL implementations might use a range of algorithms or techniques for efficient query processing: they have nothing to do with the language. The result of an XQL query is itself an XML document (check this is always true). It's useful to us in the metadata context because we can use it to extract different parts of the metadata for particular purposes. Consider the following fragment, an abbreviated, hypothetical, document in the ANZLIC DTD.

```
<anzmeta>
     <descript>
          <abstract>
          Covers Bathurst and surrounding area
          </abstract>
          <theme>
               <keyword thesaurus=ÓplacesÓ>NSW</keyword>
               <keyword>city</keyword>
          </theme>
     </descript>
     <distinfo>
          <native>
               <digform>
                    <formname>
                    Unsigned 8 bit generic binary
                    </formname>
               </digform>
          </native>
          <accconst>
          available online to the general public
          </acconst>
     </distinfo>
     <distinfo>
          <native>
               <nondig>
                    <formname>
                    ordinary hardcopy map
                    </formname>
               </nondig>
          </native>
          <accconst>
          available for purchase from gov. shops
          </acconst>
     </distinfo>
</anzmeta>
```

Suppose we want to check the data quality of the dataset described by this metadata. The DTD provides an element, `<dataqual>` for precisely this, containing sub-elements describing, for example, the accuracy, completeness and logical consistency. We get all of these as a sub-document with the XQL query `anzmeta/dataqual` and we can now check that the data meets our quality requirements. We wouldn't necessarily write these queries explicitly ourselves. They can be generated automatically by a specialised metadata query program or they can be part of a program for extracting the data itself. Let's take a more detailed look. First we will look at the way we drill down through the elements of an XML document, then at how to put in specific pattern matches.

5.6.1 Locating elements and attributes

The intense wave of activity in Web language specifications at the end of the 90s has had the considerable benefit of using similar syntax wherever possible. So the syntax for locating elements in XQL is almost identical to the pathname syntax defined for Uniform Resource Indicators. So to get the positional accuracy inside the data quality element we have simply

```
anzmeta/dataqual/posacc
```

with forward slashes (/) separating each element. We might also want to access, not the immediate children of a parent node, but simply the descendents. So we might want to pick the contact organisation nodes, which are buried quite deep in the DTD structure. Now we use a double slash operator (//): `anzmeta//cntorg`.

```
<descript>
    <abstract>
        Covers Bathurst and surrounding area
    </abstract>
    <theme>
        <keyword>NSW</keyword>
        <keyword>city</keyword>
    </theme>
</descript>
```

But what happens if the document contains repeated occurrences of an element? As you would expect, a sequence of nodes is returned in precisely the order they were in the document. However, such a list would not be a valid XML document. So the response to the query is returned wrapped inside a root element, `<xql:result>`. Suppose we have two distribution formats described in an `<anzmeta>` document. The query `distinfo` might return the XML document

```
<xql:result>
    <distinfo>
        <native>
            <digform>
            <formname>
                Unsigned 8 bit generic binary
            </formname>
            <digform>
        </native>
        <accconst>
            available online to the general public
        </acconst>
    </distinfo>
```

```
<distinfo>
     <native>
          <nondig>
          <formname>
               ordinary hardcopy map
          </formname>
          <nondig>
     </native>
     <accconst>
          available for purchase from gov. shops
     </acconst>
</distinfo>
```

```
</xql:result>
```

with descriptions of two formats, one online and the other a hardcopy format available for purchase. We have got here the immediate children (abstract) and the contents of these children. What we have actually got (as the default) is a deep return, in which we have all the children of the node. In fact a deep return is indicated by two question marks (??), whereas a shallow return is represented by just one as distinfo? returning

```
<xql:result>
     <distinfo>
     <native>
     </native>
     <accconst>
     </accconst>
     </distinfo>
     </descript>
</xql:result>
```

The reader familiar with operating systems such as UNIX might have noticed the resemblance to the path nomenclature. Similarly a few other ideas from regular and logical expressions transfer across. One useful concept, is the asterisk (*) to represent a wild card. Thus the query

```
descript/*/keyword
```

finds all keywords that are precisely the grandchildren of the descript element.
    There is just one more little twist to think about, before we move onto attributes and the text content of elements. We can select a set of nodes in a query, but return some function of this set. So cover up the solution, and see if you can guess what this query will produce: descript??//keyword|Tricky.
    What we get is the whole of the descript element (deep return) which has a keyword descendent, i.e.

```
<xql:result>
     <descript>
          <abstract>
               Covers Bathurst and surrounding area
          </abstract>
          <theme>
               <keyword>NSW</keyword>
               <keyword>city</keyword>
          </theme>
     </descript>
</xql:result>
```

So far we have discussed how to move around in and select items from the document hierarchy. There are two other useful concepts: `sequence` and `position`, but these will take us beyond the scope of the present book.

All the above tricks and techniques apply to attributes too. All we need to do is to prefix the attribute name with the at character (@). So in the following example: **@thesaurus** we get the value places returned. Attributes can occur in any order in an element and do not have sub-attributes or elements. So, the attribute component of a query will come last. Note that when we issue queries against attributes, we get strings returned. These strings do not necessarily produce a valid XML document.

5.6.2 Conditional queries

So far we have just located elements, or element sub-trees and attributes, based purely on their position within the document tree. Suppose we now want to impose conditions on exactly what we return. A condition appears in square brackets ([ ]) immediately after an element or attribute. So, `native[nondig]` returns

```
<formname>
          ordinary hardcopy map
</formname>
```

We can now add comparison expressions using simple Boolean operators, denoted by **eq** and **ne**, or simply = and !=, and **native[xxxx]**. By adding Boolean relational operators, **or** and **and** we can get really complicated queries. For instance, the expression

```
keyword[$not$ @thesaurus]
```

returns the keyword `cities`.

There is more too. We can make comparisons to integer and real numbers and a whole range of additional comparison operators are part of the XQL extensions.

5.7 WHERE TO FIND DTDS AND OTHER SPECIFICATIONS

We have seen in the preceding sections a wide variety of specification for document structure and semantics. There is still a few more to come. So, how does some given document know where to find these specifications? There are two mechanisms: a generic header at the top of the document and embedded URIs throughout.

5.7.1 Document headers

In SGML we begin with quite a complicated header block specifying a great deal of things to do with syntax used for defining SGML constructs etc. We needn't worry too much about this, since XML has tended to lock in defaults for many of these options.

The key component comes right at the beginning, the *doctype* declaration in which the DTD is given a name:

```
<!DOCTYPE myDTD [
        <!ENTITY % ISOpub PUBLIC
        "ISO 8879-1986//ENTITIES Publishing//EN">]>
```

The embedded declaration defines the *public entities*, the expressions such as   to represent a non-breaking space. In this definition we have reference to an International Standards Organisation (ISO) definition, a *public text class* (ENTITIES) and a *public text description* (Publishing) and finally after the second set of //, a *public text language code* (EN for English) (Bryan 1988).

In XML the situation is a little simpler:

```
<?xml version="1.0" encoding="UTF-8"?>
<?xml:stylesheet href="annrep99.css" type="text/css"
        charset="UTF-8"?>
```

First we have the declaration of the version number of XML and an encoding specification. The second line provides something different to the SGML framework: a specific style sheet for presenting the document. A processing system may not need to make use of this: a query agent would be interested in content rather than presentation for example.

5.8 THE FUTURE

At the time of writing, XML is developing furiously. Specifications of all kinds are on the move, while cheap or free software is becoming more readily available. SGML, although it will always be around and is mostly backwards compatible with XML, is probably on the decline. Figure 5.1 shows the current situation.

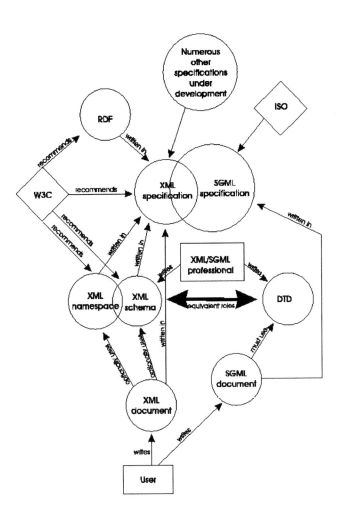

Figure 5.1. XML and SGML Components.

5.9 FURTHER READING

SGML is a much older standard than XML and there are a number of good books available. Bryan (1988) is formal and thorough and an excellent reference text. XML, like the universe, is expanding rapidly in all directions and there is really no substitute for accessing material on the web itself. The website for this book (`http://www.csu.edu.au/complexsystems/smdogis/`) has a set of current links.

Two books which discuss the broader implications of structured documentation are Alschuler (1995) and Ensign (1997).

Charles Goldfarb has been a pioneer of SGML and subsequently XML technologies and his new XML handbook (Goldfarb and Prescod 1998) is a definitive reference at the time of writing.

## CHAPTER 6

Information networks

In the preceding chapters, we have looked in detail at the mechanics of placing geographic information online. However, the real power of the Internet lies in sharing and distributing information. Geographic information is intrinsically distributed. Whether it be mines in Canada, roads in Britain, or lakes in USA, most geographic applications concern particular themes within particular regions. To compile an overview of (say) the worldwide distribution of a single theme, or else to overlay a number of themes for a given region or country, you usually have to collate information from a number of different sources.

The result is that in considering geographic information online, we need to go beyond the methods of delivering geographic information from a single Web site and look at how to coordinate geographic information that is spread across many different sites. In the following chapters, therefore, we move on to look at the issues and technology involved in doing this. The first step, which we address in this chapter, is to understand the issues involved in creating and coordinating an information network.

Chapter 6 is the first stage in integration of the tools and techniques of Chapters 1–5 into giant online systems. So far, we have seen Web technologies for spatial operations, for validating data and for markup of text material and graphics in standardised and non-proprietary form. Now we look at the concept of an information network (IN) as a synergistic union of many distinct web sites. INs require a range of supporting ideas, including :quality control and criteria for being part of the network, indexing; maintenance, security, privacy, and scalability.

In subsequent chapters, we shall examine the tools which make INs feasible. In Chapter 10 we look at the information system of the turn of the millennium: the distributed data warehouse, and how we exploit an IN to the full. In Chapter 11 we ask what new technologies and what dangers will accompany us into the new millennium.

6.1 WHAT IS AN INFORMATION NETWORK?

How do we organise information on a large scale? One approach is to start at the source and organise publishing sites into an information network. In this context an information network is a set of sites on the Internet that coordinate their activities. In particular they operate under some common framework, especially the indexing of the information that they supply.

For purposes of the present discussion we define an *Information Network* to be a group of sites on the Web that collaborate to provide information about a particular theme, subject or matter of interest. They are organisations for coordinating the development of online information. Information networks should not be confused with the communication networks that connect computers together. Just as computer networks link together computers, so information networks link together information, people and activity on particular topics.

In this chapter, we consider a broad range of information networks, which are often loosely organised and heterogeneous in nature. We consider the more tightly bound networks involved in distributed data warehouses in Chapter 10.

The Internet creates the potential to develop worldwide information systems. In a real sense, the World Wide Web itself is a giant information network. However, it fails to satisfy the second part of the definition above, namely the need for a focus on a particular theme or topic. As the World Wide Web spread, the 1990s saw a proliferation of cooperative projects to implement information networks in a number of fields. Astronomers set up worldwide networks to link star charts and to communicate about new sightings, biotechnologists set up networks, such as the European Molecular Biology Laboratory (EMBL), to provide a seamless umbrella for databases being developed at different sites.

A number of international projects have focussed on putting global resource and environmental information online. For instance, the International Organisation for Plant Information (IOPI) began developing a checklist of the world's plant species (Burdet 1992). The Species 2000 project has similar objectives (IUBS 1998). At the same time, the Biodiversity Information Network (BIN21) set up a network of sites that compiled papers and data on biodiversity on different continents. There are now many online information networks that focus on environment and resources.

In 1994, the OECD set up a Megascience Forum to promote large science projects of major international significance (Hardy 1998). The Human Genome Project was one such enterprise. Another was the proposal for a Global Biodiversity Information Facility (GBIF). The aim of GBIF is to establish

"... a common access system, Internet-based, for accessing the world's known species through some 180 global species databases ..."

Similar initiatives are also under way in primary industry. For instance in 1996 the International Union of Forestry Research Organisations established an international information network and in 1998 began work to develop a global forestry information system (IUFRO 1998).

Perhaps the most widely used information networks are those supporting popular search engines. Many search engines either farm out queries to a number of supporting databases and pool the results, or else they index source data from databases that gather primary data about sites within a restricted topic, network domain or region. The same principle has been applied in environmental data. For instance, the Australia New Zealand Land Information Council (ANZLIC) implemented an Australasian Spatial Data Directory (ASDD) to index environmental databases and data holdings (ANZLIC 2000).

6.2 WHAT CAN INFORMATION NETWORKS DO?

Perhaps the first distributed geographic information system on the Internet was the World Wide Web Index of Web sites by countries. This service, established by

CERN as the Web spread, simply sorted the Web sites registered with CERN by country. As the number of Web sites grew, this system quickly became unmanageable. In 1993, CERN began to outsource indexes for particular countries, effectively turning the service into a distributed geographic information network. The system was ultimately abandoned when control of the Web was transferred to the World Wide Web Consortium. By that stage, a combination of commercial competition and sheer growth was creating an anarchic situation in which it was impossible for any single organisation in each country to maintain an official register of sites.

Another example of an early information network with a geographic basis was the Virtual Tourist (Plewe 1997), or VT, described in Chapter 1. As we saw earlier, the VT consisted of a single world index that pointed to a hierarchy of national and regional indexes, which in turn pointed to sites providing primary data. Unlike the sites register, though, the VT's index supported geographic searching from its inception.

There are enormous benefits to gain from organising geographic information networks. As the above examples show, a geographic information network makes it possible to put together information services that no single organisation could develop and maintain. Some of these advantages include the following:

❑ *The whole is greater than the sum of its parts.* Combining different data sets makes it possible to do new things with them that could not be done individually. One example is the geographic index, with each data set covering a separate region. Another is the potential to create overlays of different kinds of data, so making possible new kinds of analysis and interpretation.
❑ *Data is updated and maintained at the source.* The organisations that gather the primary data can also publish it. This makes it easier to keep information up to date. It also overcomes many of concerns about ownership and copyright that have plagued cooperative ventures in the past.
❑ *Information networks are scalable .* That is, more and more organisations and nodes can be slotted in at different levels without the system breaking down.

The ultimate geographic information network would be a worldwide system that provided links between all kinds of geographic data at all scales (see Chapter 11). However such a system is still a long way off. Meanwhile there are enormous advantages in being able to draw together geographic information over wide areas. Gathering data has always been the most time consuming and frustrating task for GIS managers. Information networks have the potential of making widespread geographic data of many kinds available on demand.

6.3 THE ORGANISATION OF INFORMATION NETWORKS

The earliest networking projects were cooperating sites that simply provided a common interface to lists of online resources. Another common model is a *virtual library*, which consists of a central index to a set of accredited services. A more ambitious model is a *distributed data warehouse* (see Chapter 10). This system consists of a series of databases at separate sites on the Internet, with a common search facility.

An information system that is distributed over several sites (nodes) requires close coordination between the sites involved. The coordinators need to agree on the following points:

1. logical structure of the on-line information;
2. separation of function between the sites involved;
3. attribute standards for submissions (see next section);
4. protocols for submission of entries, corrections, etc.;
5. quality control criteria and procedures (see next section);
6. protocol for on-line searching of the databases;
7. protocols for "mirroring" the data sets.

For instance, an international database project might consist of agreements on the above points by a set of participating sites ("nodes"). Contributors could submit their entries to any node, and each node would either "mirror" the others or else provide online links to them.

6.4 ISSUES ASSOCIATED WITH INFORMATION NETWORKS

One advantage of information networks is that they can address directly issues that are crucial in building a reliable information system. The sites in the network operate under one of the common frameworks described above. To achieve this they need to address directly issues that are crucial in building a reliable information system. These issues include:

❑ standardisation;
❑ quality assurance;
❑ publishing model;
❑ stability of information sources;
❑ custodianship of data;
❑ legal liability and other legal matters;
❑ funding.

6.4.1 The need for standards and metadata

Coordinating and exchanging scientific information are possible only if different data sets are compatible with one another. To be reusable, data must conform to *standards*. The need for widely recognised data standards and data formats is therefore growing rapidly. Given the increasing importance of communications, new standards need to be compatible with Internet protocols.

Four main kinds of standards and conventions are used:

(a) *Information design* standards and information models describe in conceptual terms the information needs of an enterprise. All data and information are collected, stored and disseminated in the framework.

(b) *Attribute standards* define what information to collect. Some information (e.g. who, when, where and how) is essential for every data set; other information (e.g. soil pH) may be desirable but not essential.

(c) *Quality control standards* provide indicators of validity, accuracy, reliability or methodology for data fields and entries.

(d) *Interchange formats* specify how information should be laid out for distribution. The markup languages SGML and XML (see Chapter 5) provide extremely powerful, and flexible standards for formatting information for processing of all kinds. It is also extremely good for interchanging database records. The ISO standard ASN.1 tagged field format also provides a flexible protocol for defining and exchanging electronic information (Fig. 3). Software libraries now exist that provide tools to manipulate and reformat files.

6.4.1.1 Metadata

Metadata are data about data. They provide essential background information about datasets, such as what it is, when and where it was compiled, who produced it, and how it is structured. Without its accompanying metadata, a dataset is often useless. Metadata have gained considerable prominence as indexing tools, especially since the advent of large-scale repositories on the Internet. Because of the vast range of information online, the World Wide Web Consortium espoused the principle that items should be self-documenting. That is, they should contain their own metadata.

Whatever the material concerned, metadata always need to cover the basic context from which the information stems. Broadly speaking, metadata need to address these basic questions:

- ❑ HOW was the information obtained and compiled?
- ❑ WHY was the information compiled?
- ❑ WHEN was the information compiled?
- ❑ WHERE does the information refer to?
- ❑ WHO collected or compiled it?
- ❑ WHAT is the information?

For instance, the Dublin Core (DC), designed originally for online library functions, specifies a suite of fields that should be used in identifying and indexing Web documents (Weibel *et al.* 1998). An important side effect of XML (Chapter 5) is to make metadata an integral part of the organisation and formation of documents and data. The Resource Description Framework (RDF) provides a general approach to describing the nature of any item of information RDF (Lassila & Swick, 1998).

We will look at metadata standards in more detail in the following chapters, especially in Chapter 8.

6.4.2 Quality assurance

Quality is a prime concern when compiling information. Incorrect data can lead to misleading conclusions. They can also have legal implications. The aim of quality assurance is to ensure that data are valid, complete, and accurate.

To be useable, data records must be valid, accurate, and up-to-date. They must also conform to appropriate standards so that it can be merged with other data. Errors in any field of a data record are potentially serious. Important aspects of quality include the following:

- ❑   Validity and accuracy of observations;
- ❑   Accurate recording;
- ❑   Conformity to standards.

6.4.2.1 Tests on quality assurance

The most direct method of assuring quality is to trap errors at source. That is, the workers recording the original data need to be rigorous about the validity and accuracy of their results. If electronic forms are used (e.g over the World Wide Web), then two useful measures are available. The first is to eliminate the possibility of typographical errors and variants by providing selection lists, wherever possible, as we saw in Chapter 4. For instance, selecting (say) the name of a genus from a list eliminates miss-spelling (though selection errors are still possible). For free text, scripts can be used to test for missing entries and for obvious errors prior to submission. Online forms, for instance, can use (say) javascript routines to test whether a name entered is a valid taxonomic family or whether a location falls within the correct range.

Errors in methodology pose the most serious concern. For data from large institutions, the professional standing of the organisation is often deemed to guarantee reliability. However, for data obtained from private individuals, some criteria need to be applied. Publication of results is often taken as a *de facto* indicator that all data records are correct. However, this is not always true.

Everyone makes mistakes, so quality testing of data is essential. Although the methods we saw in Chapter 4 can help to guard against some kinds of errors, they do not guard against errors of fact. Perhaps the most effective way to test for errors is to build redundancy tests into data records. Redundancy makes it possible to check for consistency between related fields. For example, does the location given for a field site lie within the country indicated? Does a named species exist? If a database maintains suitable background information, then outlier tests can reveal suspect records that need to be rechecked. A record that shows (say) a plant growing at a site with lower rainfall than anywhere else needs to be checked. Either it contains important information, or it is wrong. Both sorts of checks can be automated and have been applied to many kinds of environmental data.

In principle, an appropriate indicator of quality could accompany every data field. For instance, is location given to the nearest minute of Latitude? Or degree? And how was it derived? By reference to a map? A global positioning system? Or interpolated much later from a site description? Perhaps the most important are indicators of what checks have been applied to the original records to ensure accuracy and validity.

6.4.2.2 Protocols for quality assurance

Large data repositories adopt a formal quality assurance protocol for receiving, incorporating and publishing data. Some of these protocols include testing conformity to required standards, examples of standards, publication of methodology for standards currently in use, results, and validity checks such as those described above.

Protocols for information networks are as yet less developed. One possibility is for a site to become part of an information network, it should receive an appropriate quality accreditation. For example, in manufacturing industry and software development, the ISO9000 quality mechanisms are widespread. To achieve ISO9000 accreditation requires first having adequate mechanisms in place to ensure quality of output; secondly a formal accreditation body will investigate (normally on site) the quality processes to ensure they meet international standards. Accreditation for conformance will often have to be renewed at regular time intervals.

6.4.3 The publishing model

It is important to realise that making information available online is really a form of publication. Traditionally the term "publishing" has been closely associated with books and other printed matter. However, in the modern electronic era, there are now many other formats besides print for circulating ideas. With the rise of multimedia, the distinction between print, video, audio etc. are becoming blurred. Today a more apt definition might be to describe publishing as ...

"The act of disseminating intellectual material to its intended audience" .

Although the medium and the material may differ vastly, essentially the same common process is always involved in publication (Fig. 6.1). For online publications this model makes it possible to automate many of the steps involved. The model encompasses all the stages that occur in traditional publishing, but in a somewhat more formalised form. We can summarise the steps as follows:

- ❑ *Submission* – The author submits material to the editor.
- ❑ *Acquisition* – The publisher acquires material. Here we take this to include permissions. Details of the submission are recorded and an acknowledgment is sent to the author.
- ❑ *Quality assurance* – The material is checked. Errors are referred back to the author for correction.
- ❑ *Production* – The material is prepared for publication. This stage includes copy-editing, design, typesetting, printing and binding.
- ❑ *Distribution* – The publication is shipped to stores etc. for sale. It is publicised so that people know that it is available.

Note that the above procedure is completely general. It applies to *any* kind of information, whether it be data, text, images, video, or sound. Although the details of the above process vary enormously from case to case, essentially the model is always the same. In a traditional magazine or journal, for instance, authors submit articles to an editor who records them and assesses them for quality (this may involve outside referees) and then passes them on to a production unit, which prepares the publication in its final form. An advertising or marketing unit prepares announcements when the material is ready for distribution.

This essence of the above process (as captured in Fig. 6.1) applies to publication of any kind of material, whether in a traditional context, or online. For instance, adding data to a database involves submission of data records by the custodian to the manager of the database, who runs quality assurance tests over it, arranges for it to be entered or marked up and added to the database. The delivery and announcements stages

The importance of the publication model is that it provides a systematic framework for automating many editorial and publishing functions. For instance, almost the entire submission procedure can be automated. When an author submits an item for publication (using a form upload) an automatic process can store the files, record the submission, return and acknowledgment to the author, and notify the editor. It could even carry out elementary quality checks, such as ensuring that all pertinent information has been provided, or testing the validity of embedded URLs.

When an author submits data for publication, several tasks must be performed immediately. Typically, these tasks might include:

- ❑ Assign a reference number to the submission.
- ❑ Date stamp the submission.
- ❑ Create a directory for the new material (a directory is needed because several files will always be involved).
- ❑ Write the loaded file into the directory.
- ❑ Create a registration file containing all the details of the submission.
- ❑ Add a summary and links to the relevant "incoming" queue and editorial control files for later processing.
- ❑ Send a receipt back to the author.
- ❑ Carry out preliminary checks of the information, such as checking that all fields are completed.
- ❑ Notify the editor.

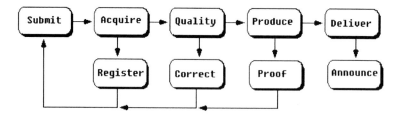

Figure 6.1. Summary of the steps involved in publishing material online. The same general pattern of steps occurs whatever the publication and whatever the type of material involved. Many of the steps involved can, and should be automated.

In general a submission will require several files to be uploaded (e.g. an article plus the figures). If the submission requires an elaborate directory system to be created, then the best approach is for authors to bundle the entire directory system and files into a single archive file (e.g. using the Unix tar facility) and submit the archive file. If circumstances do not require separate files to be uploaded as part of the submission, then the above procedure is somewhat simpler, but most of the above steps still apply. In principle there is no difference between the mechanism for submitting (say) a conference abstract, an entry to a database, or an entire book.

Another important consequence of the general publication model is that the same automatic processes that are needed for online publication will be shared by many information systems. This makes it possible to encapsulate those steps as publishing objects, complete with their own. There are now many publishing systems and packages available to create such objects. Online publishing languages, such as SLEEP, which was described in Chapter 3, also help to simplify the building of automated submission and other publishing procedures.

Finally, it should be noted that the above model really applies to publication of material on a single site. In the context of information networks, it applies only if the material can be centrally coordinated. In many contexts, a somewhat different approach – the accreditation model – needs to be applied. We look at this in more detail in Section 6.5.

6.4.4 Stability

The most frustrating problem, for users and managers alike, is that important sources of information frequently go "stale". Existing mechanisms for reporting URL changes are simply not effective. However, the solution is not to concentrate information at a single centre. An important principle is that the site that maintains a piece of information should be the principal source. Copies of (say) a dataset can become out-of-date very quickly, so it is more efficient for other sites to make links to the site that maintains a dataset, rather than take copies of it.

Publication of information online implies a measure of stability. That is, the information will be available for a long period. It will not suddenly disappear. Information networks provide frameworks for reducing, if not eliminating, the problems involved.

Perhaps the most common reason for material disappearing is reorganisation of a Web site. The material is still there, but it is in a different location within the site. Most often the problem arises because publishers fail to consider the logical structure of their site when it is first established. Another common problem is that Web sites often reflect the internal structure of the organisation that runs them. Each time the company or government department is restructured, so too the Web site needs to be changed. Publishers can avoid these problems by careful planning when the site is established, by orienting the logical structure towards users rather than owners, and by using logical names for services instead of using machine names.

A more serious problem than address changes is the actual loss of material. Information goes offline for many reasons, but especially the closing of a site or service. Most often material disappears temporarily when a network failure occurs. Duplication is the best solution.

One approach is to run a *mirror site*, that is a copy of all the material and services on another site. Some popular services have literally hundreds of mirror sites. A mirror requires that information from the primary site be copied at regular intervals. This is done most economically through a system of updates, so that only new material is copied.

Another approach is a *data archive*. Archives became popular very early in the history of the Internet, well before the World Wide Web appeared. The most common examples were anonymous FTP sites that stored public domain software and shareware. Some government sites have established online archives for various kinds of data, such as weather records, satellite images or scientific data. At the time of writing many libraries are beginning to create online archives for electronic media.

6.4.5 Data custodianship

The word custodian has two meanings: a person who *cares* for something, and a person who *controls* something. The subtle difference between these two words is important in data management. Here we deal with the two issues in turn.

Data currency is often a major problem in GIS, as well as in many other kinds of databases. For instance, city councils need to keep the names of property owners up to date for purposes of taxation, voting registration and a host of other services and issues. Likewise, business directories and tourist databases need to keep contact details up to date or businesses will suffer. The task of maintaining the currency of data is hard enough for the primary agencies that compile the information. The problem becomes compounded when a dataset is passed on to other organisations. This issue is a great concern for GIS, which are often put together by combining data from many different primary sources. An important advantage of online GIS is the potential to access primary data sets directly from the source. This means that the data updates can be incorporated into a GIS directly at their source, rather than having to wait until a new version is received and uploaded. The working versions are also the primary sources.

Attempts to share data have often foundered on the issue of control. Datasets are often valuable to the organisations that develop them. As a result, agencies are understandably reluctant to hand over to other organisations datasets that they rely on for income. Possible remedies have been tried, or recommended. These include: changing the funding models, promoting and rewarding cooperation; tying contributions from researchers to grant funding and publication of results. Creating distributed databases online provides another potential solution to the problem. Publishing their data online has allowed many agencies to retain control and management of their data whilst also making it available externally.

The problem of recouping costs still remains an issue for agencies that make their data available online, especially if the data needs to be combined with other data layers, and in automatic fashion. The problem is this. If an agency sells access to its data online, how does it charge other organisations that want to incorporate that data into other services. A good example would be (say) a geographic base layer that could serve as the base map for tourist data. At the time of writing this question is still a major issue for online data. Several approaches have been tried. Traditional models include syndication, in which an agency allows other organisations to use its data at some *pro rata* cost. Another is to pay royalties based on the rate of use of particular datasets.

The above model supposes that all of the players are aware of the way in which data is shared, and that systems are put in place to monitor this regular usage. However, the problem of micro payments becomes far more difficult in some of the automated scenarios that we shall consider in later chapters. For instance, as mentioned above the use of common namespaces in XML raises the possibility of information networks in which different sites are not even aware of one another. So what happens, for instance, if an automatic agent grabs data from another site – a one-off transaction – in the course of addressing a query? In many cases the information access may require data processing. The same arguments extend to the processing time involved.

One possibility is to pay for data by some measure of the effort required to retrieve it. This could be the actual volume of data; it could be linked to CPU cycles to process a query or some other measure of query complexity. One problem is that most current standards originally appeared before the importance of electronic commerce became apparent. Hence they do not include models to accommodate issue such as those raised here. At the time of writing, there is an urgent need for a model that enables an agent to grab pieces of data from wherever with minimal cost, or costing of just what it really uses. This could be very important for widespread applications, such as GIS access by mobile phones (see Chapter 11).

6.4.6 Legal liability

Legal liability is always a cause for concern in publishing. Perhaps the most serious concern when publishing data online is the prospect of being held liable for any damage that may result from use of that data. In some cases, the fear of litigation has itself been enough to deter organisations from releasing their data. The biggest problem arises from false, inaccurate or incomplete data and from misleading interpretations. For instance, suppose that an error in the recorded values for latitude or longitude implied that an endangered species lay squarely in the middle of an area planned for commercial development. This information could be enough to halt the enterprise at a cost of millions of dollars. The developers might then seek to litigate against all parties they held responsible, including all the persons or organisations responsible for providing the false data.

Several steps are essential to reduce the risk of litigation. There is always an implicit assumption by the public at large that published information is correct. So publishers not only need to take every step possible to ensure that information is correct, but also that users are aware of limitations. Some of these steps might include the following:

❑ Carry out quality assurance on submitted data (see earlier sections).
❑ Along with each data set provide a cover sheet that not only provides the relevant metadata, but also a clear statement that covers data quality,

limitations, and other caveats on use of the data. This is important because people tend to ignore limitations (e.g. precision) and assume that data can be used for any purpose.

❏   Along with the above caveats, provide a disclaimer that covers conditions of use and warning against responsibility. The following simple example shows the sort of detail that should be included.

> *Users of the data contained herein do so at their own risk. Although the authors and editors have made every effort to verify the correctness of the information provided here, neither they nor the publishers accept responsibility for any errors or inaccuracies that may occur. Neither do they accept liability for any consequences that may result from any use that may be made of the information.*

6.4.7 Funding

An important factor hindering many networking initiatives is how to pay for it. The economics of a single site are relatively straightforward. However, if there are many sites contributing to a particular service, then who pays for the activity? In our experience, many attempts to set up information networks fail because they entail a significant cost for the participants, without any obvious benefit. For this reason many site managers are reluctant to take part in such initiatives. This problem is particularly acute for non commercial sites, which rely on limited insititutional funds for their existence. Even when direct funding has been available to fund a network, the problem is that the large number of players means that the funding available to each site in the network is likely to be small.

In most cases the only real payoff for contributing sites is publicity. Being part of a network increases the number of pathways by which users can find your Web site (see details under accreditation below). This exposure is useful if it increases the number of users, and commercial customers.

Another tricky issue arises with commercial networks in which users pay for the information that they access. How do payments get made and how does the income get distributed? In many attempts at creating online information networks that we have seen, the agencies with data are often reluctant to contribute because they fear that the network will undermine sales of their data.

If the network is only loosely coordinated, then users might pay for individual services direct to the host site on a one-to-one basis. Problems arise where a service provided by a network makes use of elements from a number of different contributing sites.

For example, suppose that in a commercial geographic service, each map layer was maintained by a separate site and that a map building program retrieved data from each of the layers in the course of drawing a map. Presumably, the front end of the service would be maintained on a single site, which would also receive the payments from users. The contributing sites would of course, expect payment for use of their data. The question then arises as to how to organise this. There are several possible price models. For instance, the provider might need to count the number of accesses to each layer and accumulate a micro-payment each time a particular data layer is used.

In practice, the sharing of data and information along the lines described here is a matter of developing a culture in which it becomes common practice. This will happen only when the technology exists to make it practical, and pioneer services prove that it is practical.

6.5 INFORMATION NETWORKS IN PRACTICE

6.5.1 The accreditation model

The publishing model presented earlier in this chapter provides a general approach to handling primary information. However, in many cases, information networks need to compile secondary information. That is they simply link to existing information that is originally published on sites independently of the network organisation.

A particularly effective method of achieving status and visibility on the Web is the notion of endorsement, or accreditation (Green 1998). In a sense, *any* hypertext link on the web is a *de facto* endorsement. However, this is only partly true, because many sites maintain lists of "relevant" links without asserting anything about their quality. Even more so, search engines simply index sites that contain key words, without assessing quality in any way. Where accreditation differs is that the indexing site, makes a specific statement about the quality of the indexed site (Green 1995). More to the point, it excludes sites that are not considered to be of sufficient quality.

Accreditation works as follows. An organisation that monitors information quality endorses a service provided by some content provider. In practice this means providing a some form of badge or label that the endorsed site can place on its home page. Some organisations have made their reputation solely on the basis of providing an endorsement process. However, the process is a particularly effective way for organisations that already possess some form of authority to enhance and exploit their credibility in the Web environment (Green 1998).

Accreditation is particularly suited to developing networks of geographic information. Almost any kind of information can be organised geographically. Businesses are always located *somewhere* and would benefit from the ability for users to search for them by city or region, as well as thematically. Likewise, it is relevant to be able to search governments, schools and most institutions geographically. This potential richness of information means that any geographic index needs to be able to link to, and index, information from a wide range of Web sites. The accreditation model provides a convenient approach to ensuring quality of the matter that the network indexes. The exact requirement for accreditation may vary, often it is simply how prominent the site is. However, for a network that stresses quality of service, some of the most common requirements are as follows:

❑ relevance of the resource;

❑   quality of the information (accuracy, validity etc.);
❑   absence of inappropriate material, such as pornography, seditious or
     criminal information.
❑   conformance to any presentation requirements;
❑   inclusion of essential metadata;
❑   demonstrated stability of the resource, e.g.

    •   copies at mirror sites;
    •   stability of addresses (using aliases);
    •   notification of changes or closure.

❑   freedom from concerns over legal liability, copyright etc.

To test of the efficacy of accreditation, we monitored access rates to one of the services that we manage, the Guide to Australia (Green, 1993a), both before and after introducing accreditation early in 1999. The results (Fig. 6.2) clearly show that accreditation can increase the hit rate to a particular service by an order of magnitude. The reason for this is that creates many new avenues by which users can navigate to the service concerned.

Accreditation can take several forms, for example:

❑   a reference to the item in an index;
❑   a reference to an item as though it were a publication on the local site;
     providing a "badge" of approval that the publisher can include on the site
     or in the item.

6.5.1.1 Why use accreditation?

There are many advantages in accreditation for the accrediting site and organisation:

(a)   it reinforces recognition of the authority of the accrediting organisation;
(b)   it distributes the effort of developing material amongst other sites;
(c)   it encourages other sites to contribute material;
(d)   it encourages stability of contributing sites;
(e)   it helps to distribute the effort and cost of developing an information
       system;
(f)   it enables the accrediting site to impose standards and quality control on
       material published elsewhere;
(g)   it extends the number of links and references to the accrediting site;
(h)   it opens the potential for the accrediting site to "set the agenda" in the area
       concerned.

6.5.1.2 Advantages for the accredited site

Advantages for the contributor are similar to those for the accrediting organisation:

1.   increasing the credibility of the publication;
2.   extending the range of links and references to the site;
3.   public recognition;
4.   affiliation with an authoritative or high-profile organisation.

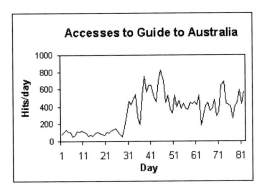

Figure 6.2. An example of the increase in Web traffic that results from accreditation of sources. Daily hits for *Guide to Australia* from 1 February to 23 April 1999. Accreditation was introduced on day 29 of the study.

There are many low-level forms of endorsement on the Internet already. Some example:

- The World Wide Web Virtual Library is simply a list of indexes that are held and managed elsewhere.
- Several sites have attempted to create a reputation for themselves simply by providing badges or "awards" to other sites. An example is the award of a badge to (say) "the 100 most popular web sites".
- Many sites that try to provide services on particular topics provide pointers to selected services elsewhere. This selection and indexing is a *de facto* form of endorsement.

By implication, accreditation implies a stamp of approval. Contributors who offer items for accreditation need to know what is expected of them. The list of requirements needs to ensure that the accredited item is suitable. On the other hand it should not be so prohibitive that it discourages anyone from making contributions. Some possible criteria include the following.

6.5.2 Examples of geographic information networks

We have already alluded elsewhere to online geographic services that are effectively information networks of one kind or another. The Internet creates the potential to develop worldwide biodiversity information systems. As the World Wide Web spread, the 1990s saw a proliferation of cooperative projects to compile biodiversity information online.

A useful outcome of networking activity has been to put a lot of biodiversity information online, such as taxonomic nomenclature and species checklists. One of the first priorities was to develop consistent reference lists of the world's species. A number of international projects have focussed on putting global biodiversity information online. For instance, the International Organisation for Plant Information (IOPI) began developing a checklist of the world's plant species (Burdet 1992). The Species 2000 project has similar objectives (IUBS 1998). At the same time, the Biodiversity Information Network (BIN21) set up a network of sites that compiled papers and data on biodiversity on different continents. There are now many online information networks that focus on environment and resources (Table 6.1).

The greatest challenges in collating biodiversity information have been human, especially legal and political issues, rather than technical problems. One outcome of the Convention on Biological Diversity was agreement on the concept of a Clearinghouse Mechanism (UNEP 1995). This scheme aimed to help countries develop their biodiversity information capacity. The longer-term goal was to enhance access to information through the notion of a system of clearing houses. These sites gather, organise and distribute biodiversity information. The greater challenge is to merge these clearing houses into a global network.

In 1994, the OECD set up a Megascience Forum to promote large science projects of major international significance (Hardy 1998). The Human Genome Project was one such enterprise. Another was the proposal for a Global Biodiversity Information Facility (GBIF). The aim of GBIF is to establish

"... a common access system, Internet-based, for accessing the world's known species through some 180 global species databases .."

Primary industries have set up similar initiatives. For instance, in 1996 the International Union of Forestry Research Organisations established an international information network and in 1998 began work to develop a global forestry information system (IUFRO 1998).

6.5.3 Further prospects for information networks

In this chapter we have focussed mostly on the practical and human issues involved in organising information networks. However, there have been many technical difficulties as well, such as coordinating updates, and developing and maintaining common indexes. We have glossed over most of these because the *ad hoc* approaches implemented in the past are now being superseded by developments in Web technology.

These new developments are changing the focus of networking activity. For instance, instead of being effectively top down organisations, planned to deal with a particular issue, networks can arise in bottom up fashion through sites adopting common approaches. For instance, the use of XML, with a common namespace has the effect of immediately creating a *de facto* network of sites dealing with a common topic area. This effect has been foreseen by the W3C, and new standards, such as the Resource Description Framework (see Chapter 8) are being developed to enhance the ability of metadata to contribute to the coordination and linking of online information resources.

In the following four chapters, we will look at these issues in more detail. Chapter 7 looks at distributed information objects, and in particular at the potential for linking online geographic information. Chapter 8 looks more closely at metadata and some of the new standards that are emerging. Chapter 9 looks in detail at the metadata standards being developed for GIS in various parts of the world. Finally in Chapter 10, we look at the concept of data warehouses and their uses.

Table 6.1. Some online biodiversity services and networks.

Organisation	*Web address* *
CIESIN	www.ciesin.org/
Clearing-House Mechanism of the Convention on Biological Diversity	www.biodiv.org/chm.html
DIVERSITAS	www.icsu.org/DIVERSITAS/
Environment Australia	www.environment.gov.au/
European Environment Information and Observation Network (EIONET)	www.eionet.eu.int/
Global Biodiversity Information Facility (GBIF)	www.gbif.org/
International Legume Database & Information Service (ILDIS)	biodiversity.soton.ac.uk/LegumeWeb
International Organization for Plant Information (IOPI)	life.csu.edu.au/iopi/
International Union of Forestry Research Organisations (IUFRO)	iufro.boku.ac.at/
Species 2000	www.species2000.org/ www.sp2000.org/
Tree of Life	phylogeny.arizona.edu/tree/ phylogeny.html
United Nations Environment Programme (UNEP)	www.unep.org/
USDA IT IS	plants.usda.gov/itis/
World Conservation Monitoring Centre (WCMC	www.wcmc.org.uk/

* All addresses should start with the prefix "http://". For example, the first entry would be http://www.ciesin.org/

Learning Resources Centre

## CHAPTER 7

Distributed objects and OpenGIS

7.1 INTRODUCTION

In this chapter we move into the transitional space between online GIS and spatial metadata. The framework we shall study, part of the OpenGIS standards from the OpenGIS consortium (OGC 2000a), provides not only a model for distributed processing of spatial data but also an elegant model for spatial metadata. So, in many ways this is a pivotal chapter. We have so far in the book talked about GIS operations and we shall go on to look in detail at the content of metadata. Here we have the architecture of a global metadata system for objects and how this would work for GIS. At the time of writing, GIS metadata is good on content description but weak on techniques for finding and exploiting it. We will come back to this in the last chapter where we look at future applications using mobile phones.

At the time of writing, vendors are working frantically to provide Web access to their GIS products. Mapinfo for example, in its MapXtreme product, provides many features via a Web interface to a remote Mapinfo server. Yet this doesn't really interest us too much here. You will find access to vendor information from the Web site accompanying this book.

This chapter involves a fair amount of low-level programming, but the user familiar with C++ or Java should find it digestible. We have two motivations for taking this close look at a widespread programming methodology. The first is the straightforward one, that these concepts and programming tools form a major part of the OpenGIS specification. The second is that it will enable us to see the complexity of GIS networks. In the (original) Web we have a simple model where documents sit on a server and are retrieved. They may contain pointers to other data. But when we want to manipulate, subset the data on remote servers or ask conditional questions of data on multiple servers, life gets a whole lot more difficult.

The CORBA specification, which we look at here, is the dominant multi-vendor standard for objects (OMG 1997), and is a product of the Object Management Group, a large international consortium.

7.2 THE STANDARDS ORGANISATIONS

Two groups play a key role in this chapter: the first is the movement behind the object-oriented technology revolution, the Object Management Group (OMG); the second is GIS specific, the Open GIS Consortium (OGC).

The OMG is a consortium of computer software vendors and other interested parties. Some 700 computing organisations got together to form the OMG, the

Object Management Group from which CORBA, the *Common Object Request Broker Architecture,* emerged.

CORBA, which is the principal standard that OMG have created, is a framework for allowing interactions of objects on different servers across the Internet. Since the role of the OMG extends far beyond that of GIS, we shall not discuss it further here, but the interested reader will find plenty of information linked to the book's Web site. The OGC is a much more specific organisation, concerned primarily with GIS. Although not driven initially by the Web, the rapid growth in the Internet has made it a powerful player.

There is a huge amount of mapping data around the world. There is also a non-trivial number of formats, data storage mechanisms, costing protocols and many kinds of impediments to *interoperability*. The Web, and distributed object frameworks, such as CORBA make it realistic to mix and match data from many sources, at least in theory. The practice is the OGC's mission.

Like many organisations drawing partners from industry, government and universities, OGC has several categories of membership from *strategic and principal* members, who have full voting rights on the various boards and committees, down to associate membership, on a sliding scale of cost. The OGC is also tightly linked through shared staff and other formal agreements to the ISO/TC211 standard which we discuss later. The OGC began in North America, but now has strong links to European initiatives and standards, with over 100 members from outside the USA. There is a specific channel through GIPSIE (GIS Interoperability Project Stimulating the Industry in Europe).

At the top of the OGC is the Management Committee, made up of representatives of principal and strategic members, which elects a board of directors. It has two other fundamental committees, the *Technical Committee* and *Interoperability Committee*. Operations of these organisations proceed, rather like the World Wide Web, through request processes: for proposals (RFP); for information (RFI); for comment (RFC); and for technology (RFT).

Responses eventually become *abstract specifications* and subsequently detailed software engineering *implementation specifications*. Several specifications are already at the implementation level, while a strategic plan documents several more proceeding well beyond the year 2000. Two important activities are the simple features specification (see Section 7.3.9) and the Web Mapping Testbed, which is the first major interoperability initiative with over 30 major corporations participating.

7.3 ONLINE OBJECTS AND THEIR METADATA: CORBA

Alongside the developments in metadata for spatial information, another impressive standardisation operation was occurring in software technology. Quite often in computing, paradigm shifts occur gradually rather than through breakthroughs. Although we can trace the invention of the Web to a single individual, its subsequent growth occurred on many fronts, in many countries through the actions of many individuals.

Another paradigm shift, not so well known outside computing circles, was the development of object-oriented technology. We saw briefly in Chapter 1 a little of objects in actions, but OOT has swept through large-scale programming like a tornado. It offers increased security, increased code re-use and a whole suite of new design methodologies which greatly improve software engineering.

Apart from new OO languages such as Java, new software design methods (Jacobson et al. 1997) and modelling languages such as UML (Booch et al. 1999), another major standardisation took place, the development of a system for managing objects in different places on computer networks, so-called distributed objects. In some ways this was a natural outgrowth of the client-server model software model (Orffali et al. 1997). This is one of the fundamental models of networked computing, where any number of clients have access to central resources maintained on a server. The distributed object model has the server provide objects to clients on demand, a bit of a simplification, but it will do for our purposes. There is a great deal of complexity "under the hood" as Orffali et al. (1997) remark.

Recapping the discussion in Chapter 1, an object is basically a heterogeneous collection of data items (possibly including other objects) and a set of methods (procedures) which act upon it. The data owned by an object is usually (really should) be accessible *only through its methods*. A *class* is a template from which to build or *instantiate* one or more objects.

In the distributed object model, we want a client to be able to operate on objects on some remote server to add or retrieve information. So we might want to query a map object for the number of people living in the region between two rivers in some county. Or we might want to get a list of all the towns in this area and do the calculation of the total ourselves if this particular specialised operation was not provided. CORBA is just one of a number of systems for distributed objects, but it has two important advantages. It is independent of the programming language used. Also, it is backed by a large consortium, the OMG.

In the early client-server models, this is fairly straightforward. We know where the data is (i.e. where the server is and, say, the database on it), and we know the language we need to extract the information. The new world is much more powerful: we don't necessarily know where the servers are, what languages they support, which the objects they manage or really not very much more than that they support the CORBA distributed object framework.

We will get onto the definition of CORBA shortly, but let's think to begin with, just what we will need.

❑ We need metadata for an object. We need to describe what it does and how you access it. We don't need to know *how* it does it, and this hiding of implementation is a key idea in object-oriented thinking. So we need to be able to describe the object's *interface* but do not need the details of its *implementation*.

❑ We need to store this information somewhere (believe it or not in the *Interface Repository* (IR), and of course, we will need mechanisms for finding IRs.

❑ We need the *implementation repository*, which stores information about the implementation of the actual object code itself.

❑   As with any global system, there is a risk of name clashes, so we need some way of assigning globally unique names or addresses to objects.
❑   We need a mechanism for finding unknown objects which will serve some specific purpose without knowing where or what they are in advance.
❑   We need a communication protocol, or, maybe something more elaborate, which will service requests to a server for objects.

As you can imagine we need some indexing and transaction mechanisms on top of the collection of objects and interfaces the server maintains. CORBA has about 16 services[1], of which we will talk about a couple of the most relevant in Section 7.3.7.

7.3.1 Interface versus implementation

One of the concepts to increase in importance in the OO revolution has been the idea of an interface. In the object model, an object owns and controls data. Other software does not access this data directly. Access is only through the object *methods*. Thus we can think the object as having an *interface* to the outside world, which is all the world sees.

How the object stores and manages its data is invisible to the outside world. In fact more than one program can implement the same interface, each doing it quite differently underneath. For example, we might want to extract the shortest distance along sealed or A-class roads between two towns. This data might be stored simply as a table for all towns on the map. Alternatively the calculation might be done on the fly by using some curve integration algorithm. We will have no way of knowing.

This idea of an interface is particularly useful when we have old data which we want to bring into contact with current software, which we refer to as *legacy* data and code. Many organisations have lots of old programs which do their job well enough and which are so important to day-to-day running that nobody dares to update them or rewrite them. In fact this reliance on and terror of touching legacy code was the driving force behind the Y2K non-crisis. However, we can *wrap* them in a modern interface, which other software can exploit.

We might want to extend their use to other applications, incorporate them in a data warehouse or simply integrate them with other data systems, say following a corporate reorganisation or takeover. Since spatial data is so costly to acquire and maintain, legacy data is a big issue in the spatial world.

Let's see how this might work with a set of maps. We have a range of maps say, dividing up the country of Scotland. Each map has spatial boundaries, a scale and maybe some ownership details. Now each of these characteristics is shared by all the individual maps each of which adds specific data. We say that each map *inherits* the parent properties, and are *subclasses* of the parent generic map (c.f.

---

[1] We say "about" somewhat hesitantly since the number of services has changed from version to version and CORBA version 4.0 is imminent. Implementation of the services has tended to lag behind everything else.

Section 1.2). We might have a specialised map that plots the distilleries for the malt whisky for which Scotland is uniquely famous. Since distilleries occur all over Scotland, from the Speyside malts in the East to the Islay malts of the North-West, we can imagine that the maps are owned by different counties and maintained on different servers. The methods of each whisky map object could print the map itself, provide information about history, tours or other interesting facts or stories.

Another map object might provide information about the Scottish Munros [2] for trekkers and mountaineers. Auxiliary information might include hazards, difficulty, date of first conquest and so forth.

In these systems we are describing, it may be that this information has been collected together from different sources, all united in a common interface. A new system might want to access both systems to provide trekking/whisky appreciation holidays[3].

To keep an interface as generic as possible we want it to be completely language independent and, preferably, conforming to an international standard protected from commercial interests. The CORBA specification gets close to this.

7.3.2 The CORBA IDL

The first building block of CORBA is the IDL, *the Interface Definition Language*. Readers who are keen programmers might recognize the similarity to C++. Although the latter was neither the first and is not universally considered to be the best object-oriented language, because it was based on the widespread C, it rapidly became the most well known. The Java programming language, which is fast becoming the *lingua franca* of Internet programming, is now seriously challenging C++ in domains outside the Internet.

In Table 7.1 we list the IDL for an interface to the Scottish maps discussed earlier. In Figure 7.1 we show the UML model of these maps.

7.3.2.1 IDL: What happens to it?

The IDL we wrote above is not program code. It can't be activated and run. It has two roles:

1. it is run through a pre-compiler to generate code which carries out the network linking functions;

2. it is stored in a globally visible repository, the *Interface Repository*, for *dynamic access*.

---

[2] Peaks over 3000 feet high.

[3] One of the tremendous advances brought about by object technology is this extensibility and reusability. It's now so easy to mix and match, as with creating a new holiday programme such as this. It has been argued that the strides made in IT underpin the substantial and sustained economic growth in the USA during the nineties.

Table 7.1. IDL for Scottish Visitor Maps.

```
module ScottishMaps//                                    Note 1
    {
    type date string$<16>$ // ISO… date
    type map; // assume defined elsewhere
    exception illegalDate{
    date illegalValue;
    string dateTemplate "YYYY-MM-DDThh:mm" //      Note 2

    +interface topoMap
    {   attribute integer sheetNumber;
         attribute string locale; //               Note 3
         void printMap(); // generic map printing function
    -}
    - interface whiskyMap:topoMap; //               Note 4
    {
         map addDistilleryToMap();
    -}
    - interface distillery
    {   +attribute boolean distilleryTours;
        attribute boolean cellarDoorSales;
        attribute string address;
        boolean open(in date);
            Is distillery open?
        + raises illegalDate;
    -}
    interface munro
    {   enum difficulty {easy, medium, hard, severe,
                extremelySevere}
        +attribute heightInFeet;
        attribute difficulty rating; //             Note 5
        +attribute integer scale;
        boolean access(in date); //                 Note 6
    -}
    interface munroMap:topomap
    {   void addMunroToMap();
    -}
}
```

1. The module name defines the namespace. So we can now uniquely refer to topomap as
    ScottishMaps::topomap
2. Dates are given according to ISO561, requiring a string of 16 characters according to a
    specific template of year, month, day and time.
3. We give the map a sheet number and a location area. This section is really just a super
    cut-down metadata set for illustrative purposes.
4. The whiskyMap class inherits topomap, since a whisky map is a topomap.
5. The rating of a Munro is a category of difficulty to get to the top. This is an
    enumerated type where, as in C or C++, we just list the possible options.
6. A lot of Scottish land has public access but is privately owned. Deer are kept and shot.
    It is extremely advisable to check if access to a particular mountain is safe!

Talking about the pre-compilation stage without detailed code analysis is tricky. Not everybody finds reading code a lot of fun; code is meant for computers to read and the best way of understanding a computational framework is to go right in and experiment with it on an actual machine. We will look a bit later on at just how we *invoke* a remote object. In other words, we have a real, active object on a server, and a virtual or proxy object in the client. The client behaves just as if it were operating on a local object. The networking is transparent.

But the client needs information about where to find the remote object; the piece of code which does this is usually referred to as a *stub*, but the client programmer doesn't pay very much attention to it. This is the first code fragment from the pre-compiler. The server needs to be able to plug the object code into the network in some way, for which it needs a *skeleton*; this is the other principal code fragment generated by the IDL pre-compiler.

7.3.2.2 Global names

Now each object needs some sort of global name, just as every machine on the Internet has a unique name. In this case the name is referred to as the *Repository ID* and, just like a URL, it is made up of a hierarchy. There are two forms, IDL and DCE. Each has three levels:

1. the first part of the name is just one of two keywords, IDL or DCE, the latter standing for Distributed Computing Environment, a standard of the *Open*

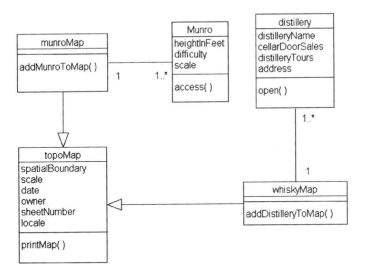

Figure 7.1. A UML Class diagram for Scottish Maps of whisky distilleries. See Table 7.1 for related IDL. Each box (representing a class of objects) provides a class name, a list of attributes, and some class operations (methods). The lines indicate relationships between different classes. See Chapter 1 for explanation of UML class diagrams.

Software Foundation (OSF 2000);
2. the second part really carries all the information – in the IDL case it is typically a URL; in the DCE case it is a unique code, referred to as a UUID (*Universal Unique Identifier*), generated by a combination of the current date, the network card address of the computer generating the code, and a counter value, finally converted to an ascii string;
3. the last part is simply a version number as in, say, 1.2. IDL allows major (1) and minor (2) versions, whereas DCE does not.

Thus we might have

```
IDL:clio.mit.csu.edu.au/terry/myobjects:1.5
```

or

```
DCE:s6ss8c8z12p4z5:3
```

These codes are generated by so called *pragma* directives to the compiler, e.g.

```
#pragma version junk 3.0 // version number for name junk
#pragma prefix "clio.mit.csu.edu.au" // default prefix
```

7.3.3 Activating objects

Activating an object is a subtle concept. Processes run on computers; objects do not, their methods do. Thus by activating an object we essentially mean instantiating it, allocating memory and other housekeeping chores and, if it's a pre-existing object, fill in the data. We also need a control model: do the objects methods run in parallel with everything else, do they sit in lightweight threads and so on? It is these loading and running issues that are the responsibility of the *object adapter*.

The first version of CORBA began with a mechanism referred to as the Basic Object Adapter. But various technical difficulties, combined with the rise in Java, led to a new mechanism, the POA, or *Portable Object Adapter*, and we will concentrate our interest on just this.

7.3.4 The Portable Object Adapter

The ORB functions as network glue. It connects clients to servers, proxy to remote objects. For very simple applications we would not have to worry about anything else. But for large systems, and GIS systems usually involve large amounts of data, there are further complications. It is not practical to keep all of the data, all the objects, online all the time. Thus we need mechanisms of moving data in and out of files or databases in a transparent fashion. This is where *object adapters* come in to play.

The first versions of CORBA used a so-called *Basic Object Adapter* (BOA), but it soon transpired that it was not a trivial software problem. In fact the specifications were sufficiently loose, that too much diversity of implementation and functionality was possible. Ultimately, the BOA was abandoned in favour of a new mechanism, the *Portable Object Adapter* (POA).

The process of retrieving a Web page is relatively straightforward: the server extracts a page from a file or database, and transfers it to the client using the appropriate (HTTP) protocol. A Web form is a little more complicated: now one or more additional programs are run to pre-process the data before returning it to the client. This chaining to other programs and consequent process creation leaves all sorts of opportunities for hackers to get into a Web server's machine and CGI programming (Chapter 3) did create major security holes in the early days of the Web.

It's not surprising, therefore, that the process of activating and exploiting objects is fraught with complications. For this reason the POA is not that simple despite its innocuous sounding name. Essentially we want to preserve this look of all-the-time-online to the prospective clients. But different systems may have different *policies* for where the objects live and how they are activated. A simple example of the difficulties might be the difference between storage in files in a hierarchical directory system and storage in a relational database. This all has to be transparent to the client.

The full details of the POA take us a little too far from our central theme, but we will take a conceptual look at two different aspects: first there is the relationship between processes and objects; secondly the different *policies* that an object adapter may use.

7.3.4.1 Object activation

What does it actually mean to activate an object? An object is not a process, a program, running in a computer. It is an assemblage of data and runnable code. A server is a process. Object activation refers more to the initialisation of the object and the registration with the CORBA system. Registration, essentially, means connecting with the POA. Now there are several ways the POA can activate objects, and a range of policies which it can use.

The options for activation are:

1. *shared server* is the most common in which each object method becomes a separate thread in the server. Suppose we have a map object from which we can add names of people living in each house shown on the map. Since people move house from time to time, we need also to be able to update the resident names. What we don't want to happen is for somebody to read data off the map while the names in a family are being changed – which could potential create an erroneous mix of family members. Thus we need careful exclusion policies, which are fundamental to database management. A thread library, such as that provided by Java, includes all the necessary control mechanisms;

2. *unshared server* creates a dedicated server for each object. This is useful
    where a dedicated object is needed, such as controlling a device such as a
    printer;
3. *server-per-method* goes to the extreme of allocating a new process for each
    method. This less commonly used. The sort of situation that might occur is to
    run some sort of house-keeping operation such as compiling access statistics.

In addition to these process mechanisms, a POA's objects may be *transient,* if
they exist only within the server, or *permanent,* if they continue to exist after the
server process terminates (i.e. they are stored somewhere).

There can be more than one POA but there is one *root* POA. Additional ones
inherit the root's generic properties but may add one or more special policies as
discussed in the next section.

7.3.4.2 POA policies

With policies we move deeper into the management of CORBA objects. In brief
outline, here are the policy options:

1. each object has to have an ID. The *ID Uniqueness policy* determines whether
    each object has a unique ID while *ID Assignment* determines whether or not
    the ID is assigned by the server or the user. Why would we need this? Well,
    the remote user is not concerned with how objects or stored, whether they are
    files, records in databases or generated in some other way. Each particular
    mechanism may require different ID mechanisms, perhaps provided by
    additional software such as a database platform;
2. *lifespan* simply determines if objects are transient or permanent;
3. the remaining policies refer to *servants.* A servant is basically the
    implementation of a CORBA object; it requires a *locator* and an *activator,*
    both sub-classes of a servant *manager.* Also required are policies on *retention*
    and *request processing*, which determine whether maps of servant/Object IDs
    are kept in an *Active Object Map* or whether requests are serviced as they
    come in.

7.3.4.3 Object activation: recap

So, we've had a whistle stop tour through some very complicated software
concepts. To the end user, most of this is transparent. To the server programmer it
can be very important. We have seen that going from metadata to effective use of
the data it describes maybe non-trivial. At the time of writing not many
implementations exist in the spatial information field, but the situation may change
rapidly, perhaps driven by the mobile phone revolution as we discussed in the last
chapter.

These concepts are important in designing server systems which are secure
and can handle large amounts of data in an efficient way. The large datasets of the
GIS user make these issues really important.

7.3.5 Enterprise Java Beans

The language Java has now gone somewhat beyond the original CORBA specifications in the development of large scale object management in the so-called *Java beans* and subsequently *enterprise Java beans* standards.

There are two huge gains in commercial productivity from object-oriented software engineering. The first comes about from encapsulation, the safe (and hopefully easy) reuse of software. The second is just beginning to take off, the building of large scale components and frameworks. Layer upon layer of software now provides easy integration of enterprise level software. The Enterprise Java Beans standard is a prime example. All of the complexities of server activation, object persistent, transaction control and so on are now hidden within the EJB framework.

However, all may not be over. Java has made progress like no other language ever. Yet it is still far from being universal. Java gets its universality by compiling to *byte codes*, which are then interpreted by Java Virtual Machines. This latter stage still tends to produce slow execution. The big changes in successive versions have made software development more complicated and not so backward compatible. But, even worse, at the time of writing, lawsuits between Microsoft and Sun are still in progress. It is possible that Microsoft may opt out of, or be forced out of, Java, which will create a tsunami in the software world.

7.3.6 CORBA services

The interaction mechanisms are all embedded within the CORBA services. The OMG has released a series of *Requests for Proposal* (RFP), beginning in 1993. For our present purposes, we will need only a small subset. The full collection as of 1999 can be grouped into several functional areas, albeit with some overlaps.

7.3.6.1 Object naming and location

The *naming service* gives objects unique names anywhere on the Internet, while the *trading service*, facilitates searching for objects out there somewhere. These are often likened to the White and Yellow Pages of a telephone directory. Two other services allow dynamic linking and naming of collections: *relationship* looks after establishing links between components which know little of each other; *properties* allows dynamic association of named values with components.

7.3.6.2 Object storage and access

The *persistence service* which renders objects available at all times indefinitely into the future; it's essentially like a transparent archiving service, usually hooked to a database, but the intricate details are of no immediate concern.

Objects may be stored in a variety of ways, in relational databases, object or hybrid databases or as simple flat files. In fact traditional GIS packages such as ARC Info may utilise different databases such as Oracle or Informix. The persistence service provides a clean, vendor and platform neutral access to objects, so that an object will appear to be online at all times. Other services are *life cycle*, for creating, deleting and moving components around on the ORB; the *collection service*, for managing collections of objects; and the *externalisation service*, for streaming (large) volumes of data in and out of a component.

7.3.6.3 Object management

Although this scenario of having globally accessible data and applications which can mix and match at will is very attractive, it brings with it some important managerial issues: integrity of data and access rights. In any database system, transaction fidelity is a core requirement and we have similar services in CORBA: *concurrency* for providing locks to prevent data corruption by more than one attempt to access the data at the same time; *transaction service* to manage the two-phase commit process of concurrent access. Synchronisation and access control require, in addition, a *time service*.

In addition to these technical problems of making sure that as many users attempt to access and update they do not damage the data, it is necessary to make sure that these users do have the necessary access rights. The *licensing service* handles measurement of usage and charging while the *security service* handles the ever present needs of authentication of users, privacy, data security and so on.

7.3.6.4 Object usage

At long last we get down to actually using these remote objects. We have a query language, similar to SQL3, in the *query service* and facilities for triggering actions based on events, controlled by the *event service*.

7.3.7 CORBA: The practicalities

It is important to realise that CORBA is a specification, and a complex one at that. It is not a software package or set of software libraries. In practice, the various CORBA implementations have all lacked some of the services. Whether full implementations will materialise, or whether some of the features will evolve out of the standard is hard to say at this stage. Java has its own set of CORBA-like and CORBA-derived distributed object libraries, but has not and probably will not, implement the full specification.

The Open GIS Consortium has released IDL for a wide range of spatial information processing tasks, which we consider shortly. Thus the opportunities are available for a widespread deployment of distributed spatial objects. But there may be a difficulty yet to be resolved. In most parts of the world, spatial data costs money, a lot of money. An e-commerce model will be needed to sell data online. We return to this issue in the last chapter.

7.3.8 The simple features specification

The CORBA IDL bindings fall into two sets: the geometry bindings and the feature bindings. We shall now look at these in turn.

7.3.8.1 Geometry bindings

Here we have the primitive vector operations you would find in most typical GIS packages. Most of these are fairly straightforward. So, for example, the *PrimeMeridianInterface* defines the prime meridian with respect to the Greenwich Prime Meridian:

```
interface PrimeMeridian : SpatialReferenceInfo {
    attribute double longtitude;
    attribute AngularUnit angular_units;
    };
```

The *SpatialReferenceInfo* interface is closely linked to the European Petroleum Survey Group (EPSG) and the Petrochemical Open Software Consortium (POSC) (EPSG 2000) and provides a set of common attributes:

```
interface SpatialReferenceInfo {
    attribute string name;
    attribute string authority;
     attribute long   code;
    attribute string alias;
    attribute string abbreviation;
    attribute string remarks;
    readonldy attribute string well_known_text;
    };
```

The terms `name`, `code`, and `alias` refer to assigned names and codes of EPSG, while for EPSG assigned data, authority is EPSG. The other items are self-explanatory, except for the last attribute which is just a textual representation of the parameters.

A comprehensive set of interfaces covers, ellipsoids, linear and angular units, coordinate systems etc.

The *SpatialReferenceSystem* interface is the parent abstract class for all spatial reference systems. This parent class gives rise to interfaces such as the *GeographicCoordinateSystem* and the *GeodeticSpatialReferenceSystem* .

7.3.9 Feature model

Features may have both spatial and non-spatial content: e.g. a town has a name and population (strings) but also coordinates, spatial extent, maps and other spatial parameters. Features as defined in the IDL interface thus consist of:

- a FeatureType
- Properties
- Geometry.

The relevant parts of the IDL (slightly simplified relative to the specification) are then

```
interface Feature {
   exception PropertyNotSet {} : // one of a number of
Property error exceptions
   exception InvalidProperty {};
   exception InvalidParams {string why;};

   readonly attribute FeatureType feature_type;

   Geometry get_geometry(in string name) raises
      (InvalidParams);

   boolean property_exists(in string name) raises
      (InvalidProperty);

   any get_property(in string name) raises (PropertyNotSet );

   void set_property(in string name, in any value) raises
      (InvalidProperty);

};
```

Properties are captured by name-value pairs with arbitrary type. The *FeatureType* is relegated to a separate interface and along with it a range of OOT *patterns* spring up. *Factories* are used to create Feature instances on the fly. Features may also be grouped into FeatureCollections. They may have a specific FeatureContainer associated with them. But they may also simply be some looser collection connected together by some property such as minimum population size of towns. Alongside collections come another standard OOT technique, the *iterator* pattern, which allows one to step through the elements of a collection without having to access its internal representation in any way. A first class reference on OOT patterns, by one of the pioneers, is Gamma (Gamma et al. 1995).

Feature collections also support a queryable interface, with the name *QueryableFeatureCollection*. The implementation details will depend on things like the nature of the database holding the feature information.

7.3.10 OGC metadata

The abstract specification for metadata in the OGC looks very promising. It is built from the ground up on an OO model. This brings with it numerous advantages:

❑ effective inheritance of parent values;

❑   minimal redundancy in repetition of information;
❑   fast and precise searching algorithms are easy to write.

At the top level, every FeatureCollection has a mandatory property, with name "Metadata" and value the ID of the metadata object. This value may be null, indicating that no metadata was recorded. Each feature within the collection may also have a metadata property but it is optional.

The Metadata Entity objects themselves are sub-classes of Metadata Sets. The Metadata Entities are themselves sub-classed to provide application specific information, such as the properties of roads.

7.4 THE GEOGRAPHIC MARKUP LANGUAGE

Another important initiative of the OGC is GML, the Geographical Markup Language. Why have yet another markup language for geographical information. Having studied XML and related standards in Chapter 5, we can now see that it will provide:

*   *implicit metadata* by having geographical meaningful tags we can do some spatial related searches based on text only. It is still difficult, although a fast moving research frontier, to do image searches based on image queries. So, for example, it is not easy at present to write search queries, such as find a fish in the given set of images. Auxiliary text markup is a way around such problems. As we have seen already, markup carries implicit metadata.
*   *a fast way of delivering geographical data to the web browser* As we saw in Chapter 4, there is an emerging standard for vector graphics on the web. So, we could mark up our map data, say, in SVG directly. Unfortunately this is not likely to be flexible enough. Browsers vary in resolution, they may be used by people with sensory disabilities, networks vary in speed, in fact there are many reasons for tailoring a web page on delivery. GML is written in XML so we can transform it easily to other XML schema with the many tools available. In fact transforming to SVG is likely to be the method of choice for most GML display purposes.

The OGC working draft, a *Request for Comment*, version 1.0 was released in December 1999, and is still at a very early stage of development. There are already a number of constructs under discussion, and it seems likely that the final recommendation will have a number of changes. Hence, we shall look at a detailed example, rather than go through the standard, step by step. Anybody interested in creating GML documents or implementations should check the web site for the latest developments.

7.4.1 Overview

A full GML specification consists of a *feature collection,* in which we have two components:

1.   a *spatial reference system*, which has its own individual DTD;
2.   a collection of *features*, each having spatial and non-spatial elements.

We will start the discussion by building the *geometry elements*, then bind them into features and finally a *feature collection.* Referring to the map (Figure 7.2), we start with the an elementary construct, the point:

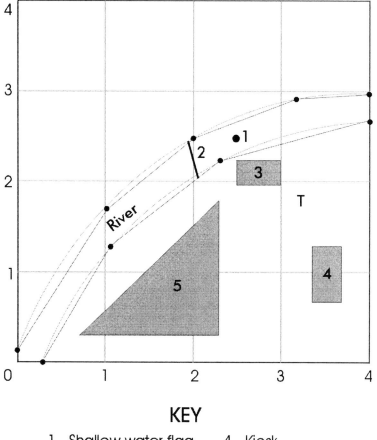

KEY

1. Shallow water flag   4. Kiosk
2. Rope across river   5. Car Park
3. Boatshed   T. Telephone

Figure 7.2. The river map defined using GML in the text.

```
<Geometry name="location" srsName="swift1701">
    <Point>
    <CList>3.2, 1.8 </CList>
    </Point>
</Geometry>
```

The point is itself made up of the fundamental element, the coordinate list or *Clist*, in this case having just one tuple. The coordinates are real-word coordinates, with a reference described in the spatial reference system, *swift1701*, about which more later.

We could make this into a feature by wrapping it in a feature element:

```
<Feature featureType="telephone" fid="1" name="Riverside
        Phone">
    <Description> The phone by the river</Description>
    <Property name="number" type="integer"> 633901013
    </Property>
    <Geometry name="location" srsName="swift1701">
        <Point>
            <CList>3.2, 1.8 </CList>
        </Point>
    </Geometry>
</Feature>
```

The feature has added a description of the feature and one of an arbitrary number of properties, in this case the number of the phone. The *fid* is simply a unique identifier. We remind the reader that this specification is not finalised and there is argument about the exact nature of the feature construct. Additionally, there is likely to be a defined thesaurus in ISO TC211 XML schema, definitions for feature types.

We have one more point element on the map, the flag in the river. Here it is as a feature:

```
<Feature featureType="flag" fid="2" name="flag02">
    <Description> Shallow water flag</Description>
    <Property name="flagCode" type="string">W</property>
      <Geometry name="location" srsName="swift1701">
        <Point>
            <CList>2.5, 2.5</CList>
        </Point>
    </Geometry>
</Feature>
```

Moving up now in complexity, we come to the rope bridge, which at the scale of the map, we just represent as a line. There is no single line construct, just a string of line segments, as a coordinate list, separated by white space:

```
<Feature featureType="structure" fid="8"
        name= "Gulliver bridge">
    <Description> Rope bridge across the river</Description>
    <Geometry name="centreline" srsName="swift1701">
        <LineString>
            <CList>2.1, 2.1 1.9 2.4</CList>
        </LineString>
    </Geometry>
</Feature>
```

We keep repeating the **srsName**; it could in fact be different for each geometry. When we come to the river, we have each bank encoded as a line string, both banks together forming a *geometry collection*. The current specification does not allow us to put the name and srsName at the collection level, which would seem a useful thing to do. Furthermore the link between features and the geometries they contain may be subject to revision:

```
<Feature featureType="flag" fid="3" name="river   Lilli">
    <Description> Delineates the river Lilli</Description>
    <Feature featureType="riverBank" fid = "4"
        name="North Bank">
        <Description> North bank of the river"</Description>
        <Geometry  name="boundary" srsName="swift1701">
            <LineString>
                <CList>
                0.0, 0.2 1.0, 1.7 2.0, 2.4 3.2, 2.8 4.0, 3.0
                </CList>
            </LineString>
        </Geometry>
    </Feature>
    <Feature featureType="riverBank" fid = "5"
            name="South Bank">
        <Description> South  bank of the river"</Description>
        <Geometry name="boundary" srsName="swift1701">
            <LineString>
                <CList>
                    0.2, 0.0  1.1, 1.3   2.3, 2.2   4.0, 2.7
                </CList>
            </LineString>
        </Geometry>
    </Feature>
</Feature>
```

We now come to the three features of the map, which all occupy areas and we represent as polygons. A polygon is simply a closed list of points.

```
<Feature featureType="carpark" fid="6" name="Car Park 1">
```

```
    <Description>
        Car Park 1 of the recreational area
    </Description>
    <Geometry  name="extent" srsName="swift1701">
        <Polygon>
            <CList>0.7, 0.3  2.3, 1.8   2.3, 0.3</CList>
        </Polygon>
    </Geometry>
</Feature>
```

The buildings are covered in a similar manner:

```
<Feature featureType="building" fid="7"
        name="Park buildings">
    <Description>
        Buildings owned by the Park authority
    </Description>
    <Feature featureType="building" fid = "4"
            name="Boat Shed">
        <Description>Boat shed for canoes </Description>
        <Geometry  name="extent" srsName="swift1701">
            <Polygon>
                <CList>
                    2.5, 2.0  3.0, 2.0  3.0, 2.2  2.5, 2.2
                </CList>
            </Polygon>
        </Geometry>
    </Feature>
    <Feature featureType="building" fid = "4" name="Kiosk">
        <Description>Refreshment kiosk</Description>
        <Geometry  name="extent" srsName="swift1701">
            <Polygon>
                <CList>
                    3.4, 0.7  3.7, 0.7  3.7, 1.2  3.4, 1.2
                </CList>
            </Polygon>
        </Geometry>
    </Feature>
</Feature>
```

We are now half-way towards completing a *FeatureCollection.* What we have to do next is to create the Spatial Reference System (SRS). There are three types of SRS: projected, based on a projection; geographic using angular coordinates on the earth's surface; and geocentric using rectangular coordinates relative to the earth's centre. Our example will use the projected system, which has the name, *swift1701,* i.e.

```
<SpatialReferenceSystem srsname=swift1701">
<Projected name="River Park">
</Projected>
```

Now we have to add information which specifies the linear units and how these related to positions on the globe. First, the units and the conversion factor to metres:

```
<LinearUnit>
    <Name>pole</Name>
    <ConversionFactor >198/39.37</ConversionFactor>
</LinearUnit>
```

The relationship to the surface of the earth comes in the geographic tag in which we specify a datum, spheroid and Prime Meridian.

```
<Geographic name="swift1701:geo">
    <Datum>
        <DatumName>Swift_Lilliput_Datum_1701</DatumName>
        <Spheroid>
            <SpheroidName>Swift 1695</SpheroidName>
            <InverseFlattening>305.112</InverseFlattening>
            <SemiMajorAxis>5944127.1</SemiMajorAxis>
        </Spheroid>
    </Datum>
    <AngularUnit>
        <Name>Decimal Degree</Name>
        <ConversionFactor>π /180</ConversionFactor>
    </AngularUnit>
    <PrimeMeridian>
        <Name>Lilliput Meridian</Name>
        <Meridian>0 0 0</Meridian>
    </PrimeMeridian>
</Geographic>
```

Finally, here is the projection itself:

```
<Projection>Swift_Conformal_Conic_Projection</Projection>
```

One more small item is left for the **FeatureCollection**, its bounding box:

```
<BoundingBox>
    <CList>0.0, 0.0 4.0, 4.0</CList>
</BoundingBox>
```

Table 7.2 gives the final assembly. The resulting map is shown in Figure 7.2. Note that this example is based on the OGC DTDs, but as we saw in Chapter 6, the future is more likely to be in XML Schema. It seems verbose! But in fact ASCII text is often a cheap storage compared to images or proprietary database formats. Remember also that markup of this kind serves multiple functions: it is self-describing through the tag names adding implicit metadata; it may be transformed into any manner of presentation formats such as SVG; it may be searched and indexed by text-based engines.

Table 7.2. A FeatureCollection Example.

```
FeatureCollection>
   <SpatialReferenceSystem srsname=swift1701">
      <Projected name="River Park">
         <LinearUnit>
            <Name>pole</Name>
            <ConversionFactor >198/39.37</ConversionFactor>
         </LinearUnit>
         <Geographic name="swift1701:geo">
         <Datum>
            <DatumName>Swift_Lilliput_Datum_1701</DatumName>
            <Spheroid>
               <SpheroidName>Swift 1695</SpheroidName>
               <InverseFlattening>305.112</InverseFlattening>
               <SemiMajorAxis>5944127.1</SemiMajorAxis>
            </Spheroid>
         </Datum>
         <AngularUnit>
            <Name>Decimal Degree</Name>
            <ConversionFactor> π/180</ConversionFactor>
         </AngularUnit>
         <PrimeMeridian>
            <Name>Lilliput Meridian</Name>
            <Meridian>0 0 0</Meridian>
         </PrimeMeridian>
         </Geographic>
         <Projection>
            Swift_Conformal_Conic_Projection
         </Projection>
      </Projected>
   </SpatialReferenceSystem>
   <BoundingBox>
      <CList>0.0, 0.0 4.0, 4.0</CList>
   </BoundingBox>


   <Feature featureType="telephone" fid="1"
         name="Riverside Phone">
      <Description> The phone by the river</Description>
      <Property name="number" type="integer">
         633901013
      </property>
      <Geometry name="location" srsName="swift1701">
         <Point>
            <CList>3.2, 1.8 </CList>
         </Point>
      </Geometry>
   </Feature>


   <Feature featureType="flag" fid="2" name="flag02">
```

```
      <Description> Shallow water flag</Description>
      <Property name="flagCode" type="string">W</Property>
      <Geometry name="location" srsName="swift1701">
         <Point>
            <CList>2.5, 2.5</CList>
         </Point>
      </Geometry>
   </Feature>


   <Feature featureType="structure" fid="8"
         name= "Gulliver bridge">
      <Description> Rope bridge across the river</Description>
      <Geometry name="centreline" srsName="swift1701">
         <LineString>
            <CList>2.1, 2.1 1.9 2.4</CList>
         </LineString>
      </Geometry>
   </Feature>


   <Feature featureType="flag" fid="3" name="river  Lilli">
      <Description> Delineates the river Lilli</Description>

      <Feature featureType="riverBank" fid = "4"
            name="North Bank">
         <Description> North bank of the river"</Description>
         <Geometry  name="boundary" srsName="swift1701">
            <LineString>
               <CList>
                  0.0, 0.2 1.0, 1.7 2.0, 2.4 3.2, 2.8 4.0, 3.0
               </CList>
            </LineString>
         </Geometry>
      </Feature>

      <Feature featureType="riverBank" fid = "5"
            name="South Bank">
         <Description> South  bank of the river"</Description>
         <Geometry name="boundary" srsName="swift1701">
            <LineString>
               <CList>
                  0.2, 0.0  1.1, 1.3   2.3, 2.2  4.0, 2.7
               </CList>
            </LineString>
         </Geometry>
      </Feature>
   </Feature>


   <Feature featureType="carpark" fid="6" name="Car Park 1">
      <Description>
```

```
        Car Park 1 of the recreational area
    </Description>
    <Geometry  name="extent"  srsName="swift1701">
      <Polygon>
         <CList>0.7, 0.3   2.3, 1.8    2.3, 0.3</CList>
      </Polygon>
    </Geometry>
  </Feature>


<Feature featureType="building"  fid="7"
       name="Park buildings">
  <Description>
    Buildings owned by the Park authority
  </Description>

  <Feature featureType="building" fid = "4"
      name="Boat Shed">
    <Description>Boat shed for canoes </Description>
    <Geometry  name="extent"  srsName="swift1701">
      <Polygon>
         <CList>
            2.5, 2.0   3.0, 2.0   3.0, 2.2   2.5, 2.2
         </CList>
      </Polygon>
    </Geometry>
  </Feature>

  <Feature featureType="building" fid = "4" name="Kiosk">
    <Description>Refreshment kiosk</Description>
    <Geometry  name="extent"  srsName="swift1701">
      <Polygon>
         <CList>
            3.4, 0.7   3.7, 0.7   3.7, 1.2   3.4, 1.2
         </CList>
      </Polygon>
    </Geometry>
  </Feature>
</Feature>


</FeatureCollection>
```

CHAPTER 8

Metadata on the Web

8.1 INTRODUCTION

The World Wide Web has grown at an enormous rate. By the middle of the year 2000, there were around 50 millions sites worldwide, some 200 million forecast by the end of 2001. In fact the growth of the Web is strongly analogous to the growth in connectivity of a random graph. This phenomenon, first observed by Erdos and Renyi (1960), underlies many properties of complex interactive systems, as elsewhere demonstrated by David Green (Bossomaier & Green 2000).

In Figures 8.1, 8.2, you can see the growth in the *connectivity* (the fraction of nodes connected together in one big component) as a function of the number of connections. As you can see at a quite small number of possible connections a *connectivity avalanche* occurs. So it is with the Web.

This unprecedented growth has brought various problems along with it:

- ❑ it's getting more and more difficult to find anything; currently the biggest search engines are each, individually, indexing less than 20%;
- ❑ some material may be offensive, yet whole scale censorship is undesirable;
- ❑ the authenticity of material may be questionable;
- ❑ it may be hard to decide on the quality or accuracy of web pages;
- ❑ personal data may be collected and used in ways not desired by the user.

Thus there is an urgent need for ways of describing web sites. This in turn creates an urgent need for metadata.

Figure 8.1. Increasing connectivity shown during the formation of a random graph, as edges are added to join pairs of vertices. Notice the formation of groups (connected subgraphs) that grow and join as new edges are added.

The HyperText Markup Language (HTML) has had, since the early days, a META tag. This tag is still the only universally recognised source of metadata. But it has become hopelessly overloaded and several new directions have emerged:

❑ the *Dublin Core* workshops have generated a full set of bibliographic tags, which assign properties such as author, creation date to web pages;

❑ PICS (Platform for Internet Content Selection) arose to label and rate content in specific ways, primarily for the use of parents and teachers;

❑ XML is growing rapidly as a more powerful and extensible alternative to HTML which is potentially self-describing;

❑ the World Wide Web Consortium recommmended in February 1999 the *Resource Description Framework* (RDF) which is a powerful structure for creating metadata for a wide variety of applications.

Our main concern in this chapter is not specific metadata tag sets, such as Dublin Core, but the more general issue of metadata paradigms. Hence our focus will be on RDF. The RDF working party had as part of its brief that it subsume PICS, so we shall briefly consider PICS first. The XML issue is more complex. XML is like a protégé who outpaces the skills of his master. Although it began as a simplified form of SGML (Chapter 5), it is now taking on a much broader role, and taking on structural characteristics which go beyond the original DTD concept. XML frequently provides implicit metadata and is the language of RDF. At the time of writing this overlap is still under discussion.

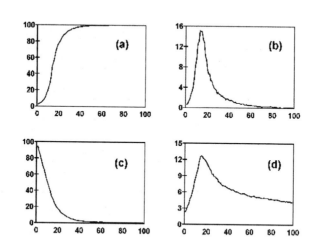

Figure 8.2. Critical changes in connectivity of a random graph (cf. Fig. 8.1) as the number of edges increases. (a) The size of the largest group (i.e. connected subgraph). (b) Standard deviation in the size of the largest group. (c) The number of disjoint groups. (d) Traversal time for the largest group (after Green 1993b).

We introduced XML in Chapter 5, but in Section 8.5 we discuss the issue of namespaces. This is the mechanism of recording metadata tags for the many possible applications to which RDF might be applied.

Having cleared the decks of preliminary material, we now look at the structure of RDF, its data model, its syntax and the *schema* mechanism through which data properties receive their definition.

We conclude by considering briefly a couple of core initiatives, still in the development phase at the time of writing: P3P (Platform for Privacy Preferences Project) and the ongoing work on digital signatures and authentication.

8.2 THE DUBLIN CORE

The term Dublin Core (DC), perhaps surprisingly, refers not to Dublin in Ireland but to Dublin, Ohio, which is home of the OCLC, the *Online Computer Library Center* and the Dublin Core directorate. The DCMI (*Dublin Core Metadata Initiative*) began in 1995 and the first Dublin Core Workshop was held there, and subsequent workshops have been held around the world including Warwick in the UK (source of the *Warwick Framework* for metadata design principles) and Canberra, Australia. Unlike the rest of this chapter, the Dublin Core is not about methods and techniques, but is a set of metadata requirements, derived from a library perspective. It is about *semantics,* rather than structure or syntax and aimed at facilitating *resource discovery* on the Internet. As such, DCMI has co-evolved with the structure and syntax mechanisms, XML and RDF.

8.2.1 Specification of the DC Elements

The reader will soon become aware of the huge range of possible sets of metadata required for different *applications*, and in this chapter we are primarily concerned with how we describe metadata elements. But the relative simplicity of the DC specification means that a simple description mechanism is adequate. Each element has 10 attributes. The first six of these do not change for any of the elements in the current version. Table 8.1 lists them; they contain few surprises.

Table 8.1. Fixed attributes of DC elements.

Attribute	*Value*
Version	1.1
Registration Authority	Dublin Core Metadata Initiative
Language	en[1]
Obligation	Optional
Datatype	Character string
Maximum Occurrence	Unlimited

---

[1] English, according to ISO639.

Table 8.2 gives the four variable elements, again fairly obvious in intent, to define the *Creator* element.

Table 8.2. Variable attributes of Dublin Core elements.

Attribute	Value
Name	Creator
Identifier	Creator
Definition	The agent(s) responsible for the content of the resource
Comment	Examples of a creator include

In most cases the name and identifier are the same, but occasionally the name is a somewhat more extended description than the (unique) identifier.

8.2.1.1 The Dublin Core elements

Having cleared the structural details of the definitions out of the way in the previous section, we can now look at the 15 DC elements themselves, given in Table 8.3 in simplified form.

Table 8.3 The Dublin Core Elements

Name	Identifier	Notes
Title	Title	
Creator	Creator	
Subject and Keywords	Subject	
Description	Description	
Publisher	Publisher	
Contributor	Contributor	
Date	Date	Use of ISO8601 is recommended, i.e. YYYY-MM-DD
Resource Type	Resource	Nature or genre of the resource
Format	Format	Physical or digital manifestation
Resource Identifier	Identifier	URL, ISBN etc.
Source	Source	
Language	Language	Use ISO639 two letter language codes with optional two digit language code from ISO3166, e.g. "en-uk"
Relation	Relation	
Coverage	Coverage	May be temporal, spatial or administrative
Rights Management	Rights	Copyright etc.

Note that some of these descriptors are not complete in themselves but require other standards or *formal* identification systems. In the case of URL, ISBN, these are in widespread use. But for other elements, such as Relation, there is no obvious standard to use.

We see in the next section, how we *can* incorporate these elements into an HTML document. There is no standard for this, but an Internet Society memo exists (ISOC 2000).

8.2.2 The HTML META Tag

An HTML document consists to two parts: the *head* and the *body*. The head does not contain information for display in the browser (although the title element often appears on the banner of the frame surrounding the web page). In principle the browser can just retrieve the head of a page and determine whether it wants the full page. One immediate use here is to determine if the page has been updated since the browser last accessed it, although this usage has tended to be circumvented by other techniques. However, the head can contain information about what is in the page, the metadata, which can help determine if the user actually wants to download it.

When the web first started, few people can have realised just how much and how rapidly it would grow. Thus the provisions for metadata were fairly limited, in fact, to just one empty tag[2] <meta>. Each occurrence of the tag contains a property/value pair. Fortunately, multiple metadata tags can be included and it is sometimes used to carry the whole Dublin Core!

The <meta> tag has several attributes:

name: refers to the name of the metadata property, such as the author of a document;

content: is the information specified in the name attribute, such as the author's name, i.e. the value of the property;

http-equiv: this gets a little bit more tricky and relates to the protocol for retrieving the page; we go into a little bit more detail below;

scheme: interpreting the value of the property defined in a name/content pair may require access to external information of some kind referred to as a scheme;

lang: there are some other, global attributes, which can be included, which are not particularly important to our needs; one such is the language of the document.

Here is a straightforward example[3]:

```
<META name="author" content="T. Bossomaier" lang="en">
```

The attribute *http-equiv* is an alternative to the *name* attribute. What it

---

[2]   i.e. no end tag.

[3]   Note that this is an SGML (HTML DTD) construct, where the end tag is omitted. In XML this would not be permissible and the tag would end with />.

does is to create a special header in the form of the HTTP protocol used to fetch pages from web servers, but the details would be a digression.

A frequent use of the `<meta>` tag is to provide keywords for the page, to facilitate the task of search engines, in the example below indicating the page contains map information in two different languages:

```
<META name="keywords" content="maps, spatial, OS, UK"
    lang="en">
<META name="keywords" content="cartes, spatial, france"
    lang="fr">
```

Here's a more complicated Dublin Core example

```
<META name="DC.Creator"    content="Terry Bossomaier" >
<META name="DC.Creator"    content="David Green">
<META name="DC.Title"      content="Spatial Metadata and Onlne
    GIS" lang="en">
<META name="DC.Publisher"  content="Taylor and Francis">
<META name="DC.Language"   content="en" scheme="ISO639">
```

Note that we capitalise the element after the prefix and the elements may appear in any order. We have not included all the DC elements in this example, nor do we have to. The prefix DC, of course, refers to Dublin Core. But how does the Web page make this absolutely clear? It uses the `<LINK>` element.

```
<LINK REL="schema.DC" HREF="http://purl.org/DC/elements/1.0">
```

Note that DC is just one of many possible prefixes, which could follow the schema keyword in the REL attribute.

There is a little complication here with language which might be confusing. In the example above, we have referred to a scheme, in fact an ISO standard, which defines the language abbreviation. Now consider the following:

```
<META name="DC.Creator" content="Tomasi di Lampedusa">
<META name="DC.Title" content="Il Gattopardo" lang="it">
<META name="DC.Title" content="The Leopard" lang="en">
<META name="DC.Language" content="it" scheme="ISO639">
<META name="DC.Source" content="ISBN 88-07-80416-6">
<META name="DC.Type" content="novel">
```

This script describes an online version of a novel by Lampedusa (it is distinguished from Visconti's superb film of the book, which could in principle be embedded within a web page, by the DC.Type value). The resource itself is in Italian, as indicated by the DC.Language attribute value. But there are two titles, one in English, the other in Italian, indicated by lang attributes within each META element. Finally, this online version was derived from an original printed novel, which we describe by its ISBN number.

Now a couple of issues should become apparent here, which will become important later. First, this is a *flat* format, just an unstructured list of property-value pairs. So, as the metadata gets more extensive and complicated it is a bit difficult for a human reader to digest. This may not be so much of a problem for a machine reading the data, but it can be a problem for a human author creating it.

Secondly, on a web site or some other document collection there is likely to be a lot of repetition. Take for example the Web site of a computing department in a university. There are a number of generic properties for a university: it carries out teaching and research and gives degrees. A university will have several faculties, arts, science, health and so on. A search engine may wish to exploit this knowledge in its search, but obviously we don't want to repeat the higher level (university, faculty) information at the department level. Repetition risks errors; it increases download times and increases the workload for authors. So we need some sort of hierarchy mechanism, where we can inherit metadata from above. PICS, which we consider shortly in Section 8.4, makes a start in this direction.

Dublin Core was in many ways a library initiative and is well known amongst librarians and archivists. But metadata is only of any use if people use it and if everybody agrees on what the terms mean. We have seen that Dublin Core has a simple mechanism for standardising the description of its elements. But it's very general, as befits a minimalist description. As we want to provide more complex descriptions, we hit the problem of advertising. How do we share our metadata. PICS provides a mechanism.

8.2.3 Profiles and schemes

The `<HEAD>` element may contain a profile attribute, which specifies a set of definitions or other material relating to the meta tag properties. For example the profile might specify the Dublin Core. So, what goes into a profile? Apart from page specific material, it could contain links to other profiles at a more general level. To make this work efficiently, the browser or search engine needs to be able to cache profile information to avoid continually downloading it for each page.

It should perhaps be clear already, that the HTML `<meta>` tag is being asked to do a lot of work! Worse still, it has been the subject of abuse: keywords crammed into the document head of an HTML document, typically in the meta tag itself, are used to fool search engines into selecting a page, possibly independent of its actual content. Thus some search engines now ignore the meta tag completely.

8.3 PICS: PLATFORM FOR INTERNET CONTENT SELECTION

PICS began as the need to protect children from unpleasant material, mainly pornography, on the Web. Although there are politicians who see censorship as necessary, many others feel that the Internet should have no restrictions provided nobody suffers unintentionally. Hence the idea behind PICS was that sites would either carry label data, or their URLs would be listed in label bureaus, which would define their content. Browsers could then be configured not to accept data with particular labels. An underlying requirement was that PICS should be easy to use: parents and teachers would be able to use it effectively to block site access.

Voluntary censorship can be quite successful: as a manager of a pornographic site, you do not want to be closed down for the interests of the majority. Being able to restrict your material to your specific clientele is quite satisfactory. On the other hand, some sites with a strong marketing focus, might be not so willing to self-regulate. Thus we need an additional mechanism of third party rating. The academic world, for example, relies very strongly on peer group refereeing. Respectable journals contain papers which have been scrutinised by experts in the field before publication goes ahead if at all. The PICS label bureau model looks after this[4].

The W3C web site `<http://www.w3c.org/pics>` has full lists of documentation, mailing lists, media commentaries etc. The PICS working groups have finished. There are three technical recommendations:

❑ *service descriptions,* which specify how to describe a rating service' vocabulary; rating services and systems became a W3C recommendation on October 31, 1996.
❑ *label format and distribution*, which deals with the details of the labels themselves and how to distribute them to interested parties; label distribution and syntax also became a recommendation on October 31, 1995
❑ *PICSRules*, which is an interchange format for the filtering rule sets to facilitate installation or send them to servers; the rules specification became a W3C recommendation somewhat later on December 29, 1997.

An additional recommendation is proposed on *signed labels.* The development of this recommendation is presumably intwined with the work on digital signatures themselves[5].

Since the PICS format is subsumed by RDF, but it's interesting to look at some of the concepts. One of the perennial difficulties of a large self-organising, system like the web, with no central control, is the need to maintain backwards compatibility. So, although a new site might use RDF, PICS, as a W3C recommendation, should still be usable way into the future.

---

[4] There is a still unresolved issue here. If labels are kept on a site or within a page, then the network impact will be negligible. But if we have bureaus carrying labels for lots of web pages or sites, then they potentially become network hotspots. Obviously, label bureaus need to be mirrored, but, as they grow in popularity, ensuring fast access may become very difficult.

[5] Part of the specification includes details of recording faithfully a check on the veracity of a document. The so-called MD5 message digest serves this purpose. It then has to be signed, and therein lies an ongoing tale of control, part patent, part government. Garfinkel (1995) gives a readable account, but see also Section 8.4.4 on XML signatures.

8.3.1 PICS rating systems and services

To provide a rating for a site, we need a documented system for the rating itself and a service which provides the appropriate ratings on demand. The definition of the labels uses a lisp-like syntax as shown in the example in Figure 8.3. Note that semi-colons introduce comments and the note numbers are as follows:

1. defines URLs for the rating system and the rating service;
2. defines the type of paper which may be a tutorial or a paper reporting new research;
3. although we have specified numerical values for the different paper types, in this case these numbers are like an enumeration type in languages such as C; the values have no ordering importance, indicated by the unordered clause;
4. papers may be invited in which case they may have no referees; some categories, such as extended reviews, will have fewer referees than new research; hence the number of referees is specified as a range of from 0 to 3;
5. online journals have facilitated a whole new range of refereeing mechanisms; in this hypothetical case, readers may query the current status of a paper, possibly with access to it at pre-publication stages; in this case the labels may

```
((PICS-version 1.1)
(rating-system "http://www.gcf.org/ratings")
      (rating-service  "http://www.gcf.org/ratings");
      (name "OGIS Publication Ratings")
      (description "Describes the categories of papers in
            OGIS, the Journal of Online GIS and their status
            in the refereeing and publishing process")
      (category
      (transmit-as "type");              Note 2
            (name "Publication type")
            (label (name "research paper" (value 0))
            (label (name "tutorial" (value 1))
            (label-only)
            (unordered true));           Note 3
      (category
      (transmit-as "referees");          Note 4
            (name "referee count") value 0)
            (integer)
            (min 0)
            (max 3))
      (category
      (transmit-as "status")
            (name "Publication status");     Note 5
            (label (name "being refereed") (value 0)
            (label (name "accepted") (value 1)
            (label (name "under revision") (value 2)
            (label (name "published") (value 3))
```

Figure 8.3. Simple example of definition of PICS rating service.

have a function of excluding particular categories of reader, perhaps only a select group of people will be able to access the paper before it reaches published status.

8.3.2 Creating labels

The labels themselves have a similar lisp-like syntax. There is a range of different options available for describing the labels, which we will not go into in detail, but in the example (Figure 8.4) they are pretty much self-explanatory.

First we have a date on which the ratings were made. Fred's paper is a research paper which will be refereed by three people. It is currently under revision. This rating expires at the end of the year, which is the timetable over which Fred is expected to complete the revision.

Jill's paper is a review, which goes out to one referee. It has already been published and, as a result, this label has no expiry date set.

We should stress, that as with the other specifications through which we have taken a whistle-stop tour, we have left a lot of formal detail out. The interested reader can pursue a number of excellent references on the web site as well as the specifications themselves.

8.3.3 Label distribution

How are labels distributed? There are two main approaches:

1. embed the labels in the HTTP header stream; this requires a compliant server;
2. use the META tag; this has the disadvantage of requiring information in all pages and being applicable only to the HTML pages themselves not to other formats (images etc.). Some browsers will move up the document tree looking for generic labels if none are found in a specific page.

```
(PICS-1.1 "http://smdogis.vir/v2.0"
labels
    on "2000-10-10T14:00
        for "http://smodgis.vir/freds-paper.html"
        until "2000-12-31T23:59:00"
        by "Editor 1"
        ratings (type 0 referees 3 status 2)
        for "http://smodgis.vir/Jills-paper.html"
        by "Editor 2"
        ratings (type 1 referees 1 status 3))
```

Figure 8.4. Simple PICS labels.

```
(PICSRule-1.1
(serviceinfo (
     "http://smdogis.vir/v2.0"
     shortname "ogis"
     bureauURL "http:///bureau.smdogis.vir/ratings"
     UseEmbedded "N")
     Policy (RejectIf "(  (ogis.status <=2) and
                        (ogis.type > 0))")
Policy (AcceptIf "otherwise")
)
```

Figure 8.5. Application of PICS Rules.

An example of the use of the meta tag would be:

```
<META http-equiv="PICS-Label" content='
   (PICS-1.1 "http://clio.mit.csu.edu.au/sit-labels"
   labels on "1999.09.09T09:09-0000"
   until "2000.12.31T23:59:0000"
   for "http://clio.mit.csu.edu.au/dangerous/page.html"
   ratings (violence 10))'>
```

The University of Chicago provides a web based PICS Application Incubator which will guide you through the steps of creating, distributing and using labels.

8.3.4 Applying PICS rules

Suppose we want to restrict access of our browser to research papers which have already been accepted or published. Figure 8.5 shows how we do it.

8.4 THE RESOURCE DESCRIPTION FRAMEWORK

RDF has three distinct components. First, we have the RDF data model, which describes in a graphical notation the relationship between document components. Secondly we have the serialisation of the model and a specific grammar. The grammar tokens are just tokens at this stage. Their meaning and interrelationships are covered by the third component, the RDF schema. The data model and syntax became a recommendation of the World Wide Web Consortium (W3C 1999) in February 1999 and the schemas became a recommendation a month later.

RDF has utilised concepts and developments from several different areas, as illustrated in Figure 8.6. In general this is helpful but it can cause some confusion. Extensive use of object-oriented programming makes the structure easy to understand. But the term object also appears in the description of statements and the new term schema is used to describe an annotated class hierarchy. We will try to point out these sources of confusion as we go.

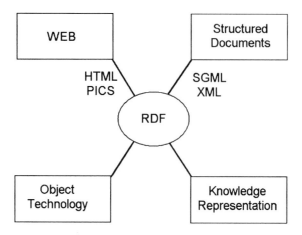

Figure 8.6. Concepts and technologies that play a part in the Resource Description Framework (RDF).

One of the key features to the success of object-oriented programming has been in software reuse. Exactly the same philosophy underlies the object-like model of RDF – reuse of metadata, particularly via inheritance mechanisms. So, for example, generic GIS data might be qualified by individual country-specific sub-classes.

8.4.1 The labelled digraph model

A convenient way of representing the data model is via labelled digraphs. There are two components:

1. *the resource* modelled by an ellipse;
2. *a property* modelled by an arc directed to
3. *the property value*, which may be a literal, represented by a rectangle, or a further resource.

The combination of resource, property and value may be thought of as a *statement* in which the resource is the subject, the property the predicate and the value the object. Note that this is one of the areas where confusion of the use of object may occur.

Consider the statement "Lilliput NMA is the owner of the Property Boundary Dataset." This situation is represented graphically in Figure 8.7.

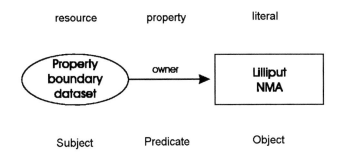

Figure 8.7. Representing ownership of a resource in the labelled digraph model.

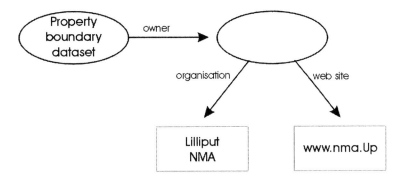

Figure 8.8. Representing joint ownership in the labelled digraph model.

Where the object concerned is a further resource, we can see that the second resource may be either labelled or unlabelled, as shown in Figure 8.8.

8.4.2 The container model

It is perfectly reasonable to have repeated resources associated with a single subject. But sometimes we would like the set of resources to have an identity in its own right. We would then put them into a container class, known as *bag*. Thus the staff who designed and do a lot of the teaching in the Bachelor of Spatial Information Systems course, BSIS, might be referred to as an entity, with particular meetings etc. Thus we would have on the web site describing the course:

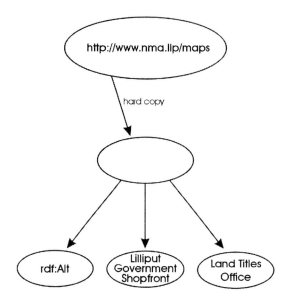

Figure 8.9. An example of the RDF container model for a resource.

```
<rdf:RDF xmlns:sit="http://clio/sit-ns">
<<rdf:description about="/bsis-staff">
     <csu:staff>
           <rdf:bag>
                  <rdf:li resource="/bsis/Bill">
                  <rdf:li resource="/bsis/Kate">
                  <rdf:li resource="/bsis/Xihua">
           </rdf:bag>
     </csu:staff>
</rdf:description>
```

The syntax here is self-explanatory, the individual bag elements making use of the HTML list element.

The **alt** container element expresses an alternative. For example hard copy maps might be obtainable from several sources:

```
<rdf:RDF xmlns:lpi="http://lpi-ns.vir">
<rdf:description about="nsw-maps.vir/central-west">
     <nsw:mapHardCopy>
     <rdf:alt>
           <rdf:li resource="NSW Government Shopfronts">
           <rdf:li resource="Land and Property Information">
     </rdf:alt>
     </nsw:mapHardCopy>
</rdf:description>
```

The syntax here is again self-explanatory. However, the serialisation syntax in XML does not make clear the notion of a type. Thus in Figure 8.9, the empty ellipse is a type. It does not have the same power of the type concept in object-oriented software technology.

The final container model is a *sequence*, for which the syntax and diagrams are exactly the same! For example, in NSW, the registering of a car involves three distinct operations, visiting three different places if one does not trust the postal service: testing the car at an approved testing station; purchasing third party insurance, the green slip; taking a test certificate and green slip to the registration office. With digital signatures, not yet legally binding in NSW, but acceptably in

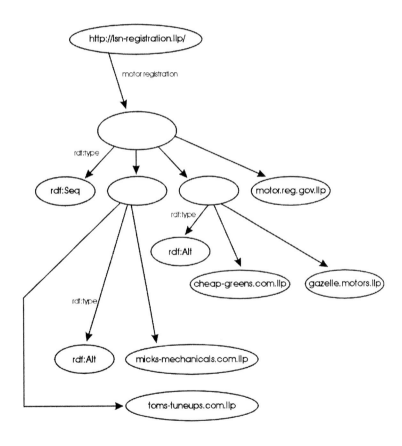

Figure 8.10. An RDF sequence.

Lilliput's Network State, we need only one visit to the garage. The rest happens over the Web. We select a garage, from which we receive a test certificate. An MD5 digest is signed by the garage using its private key. The signed digest is now entered into a second site, which is one of a number of insurance providers. A second MD5 signed digest is returned and the two[6] are now entered into the third and last government web site where the certification is obtained. Figure 8.10 shows this sequence of events and the RDF serialisation is:

```
<rdf:RDF xmlns:lpi="http://www.lsn.gov.ll/ns">
<<rdf:description about="lsn-registration.l/central-west">
     <lns:motorRegistration>
     <rdf:seq>
     <rdf:li>
<rdf:alt>
     <rdf:li resource="http://www.micks-mechanicals.com.ll">
     <rdf:li resource="http://www.toms-tuneups.com.ll">
</rdf:alt>
     </rdf:li>
     <rdf:li>
     <rdf:alt>www.cheap-greens.com.ll">
     <rdf:li resource="gazelle-motorins.com.ll">
     </rdf:alt>
       </rdf:li>
<rdf:li resource="http://www.motor.lsn.gov.ll">
<rdf:seq>
</lsn:motorRegistration>
</rdf:description>
```

A processing agent (maybe belonging to your financial advisor) could process this description and carry out all the steps on your behalf. It would have your own criteria in choosing which of the alternatives, perhaps, say always choosing the cheapest option, and could follow through this entire sequence autonomously. Your registration sticker is printed for you at the garage, almost the instant you enter your digital signature for transactions to proceed.

8.4.3 The formal RDF model

The formal model as described in (Lassila and Swick 1998) has eleven features, but some of these concern statements about statements, a process referred to as *reification*. Ignoring reified statements, which we do not have the space to discuss, leaves us with:

1. there is a set called *resources*;
2. there is a set called *literals*;
3. there is subset of resources called *properties*;

---

[6] In principle, we do not actually need the original test certification as this was a prerequisite for obtaining the insurance.

4. there is a set called *statements*; each element (statement) is a triple {pred,sub,obj}, where pred is a property, sub is a resource and obj is either a resource or a literal;
5. there is an element of properties known as *RDF:type*;
6. statements of the form {RDF:type, sub, obj} require that obj be a member of resources;
7. there are three elements of resources, which are not properties, known as RDF:bag, RDF:Seq and RDF:Alt;
8. there is a subset of properties called Ord, whose elements are referred to as RDF:_1, RDF:_2, RDF:_3 ... etc.

8.4.4 The XML syntax for the data model

RDF models have to be readable as text, for which we need a so-called *serialisation syntax*. This takes two forms: the first or basic syntax is the more straightforward, comprehensive and verbose; the second is the *abbreviated form* which handles a subset of statements in a more concise way. RDF interpreters should be able to handle arbitrary mixtures of the two.

So, let's look at a simple example.

```
<rdf:RDF>
xmlns:rdf="http://www.w3.org/1999/02/22-rdf-syntax-ns"
xmlns:cats="http://schemas.org/cats"
  <rdf:Description
  about="http:///www.cats-of-the-world.org/
  SuperCats/Garfield.html">
  <cats:Creator>Garfield</cats:Creator>
  </rdf:Description>
</rdf:RDF>
```

First, we specify the location of the RDF namespace. Next we specify a namespace *cats*. Now we define the metadata for a page about Garfield. The description element is the essential element. Garfield created his own page and is given as the creator. In fact we could have used Dublin Core to define the page creator too, but the beauty of the namespace framework is that we can use the term in a different way if we so desire.

8.4.4.1 The abbreviated syntax

There are three strategies for abbreviation:

1. instead of embedded description elements we put all the attributes into a single description providing there is no name clash;
2. we make embedded descriptions resource attributes;
3. the type property becomes an element name directly.

8.4.5 Properties of container elements

Sometimes we want to make a statement about all the pages in a container, not about the container itself. Consider the following set of pages denoting maps in Lilliput.

```
<rdf:bag ID=''maps''>
      <rdf:li   resource="http://landinfo.gov.ll/redMountains">
      <rdf:li$ resource="http://landinfo.gov.ll/forestLake">
</rdf:bag>
<rdf:Description aboutEach=''#maps''>
<llm:mapMaker>Gulliver</llm:mapMaker>
</rdf:Description>
<rdf:Description about=''\#maps''>
<llm:custodian>Lilliputian Mapping Authority</llm:custodian>
</rdf:Description>
```

The effect here is to define Gulliver as the map maker for *each* map in the set using the *aboutEach* construction, while the custodian of the *collection* is the Lilliputian Mapping Authority.

CHAPTER 9

Metadata standards

In previous chapters, we have looked in detail at how we can use metadata to index online resources, especially spatial information. However, metadata will have only limited value if everyone uses different terms. To be useful in practice, standard metadata vocabularies are necessary. In particular, to enable the mixing of spatial data between sites and applications, requires standards for spatial metadata. At the time of writing these standards are emerging through national and international organisations.

In this chapter, we look at some of the existing spatial metadata standards. As a case study, we will look at the Australasian standards in detail. This will include a summary of the ANZLIC DTD. We will also look at some activities spatial metadata standards in other parts of the world, specifically North America and Europe.

9.1 AUSTRALASIAN STANDARDS

9.1.1 ANZLIC background

ANZLIC is the Australia New Zealand Land Information Council and is a composite of a number of different groups. It has been working on metadata. In 1994 it adopted a policy on the transfer of metadata. ANZLIC formed a working group in April 1995 to develop a metadata framework and began work on the spatial data infrastructure for Australia. Throughout the process was a consultative one, leading to shared ownership of the standards. This contrasts with the position in other countries where standards imposed from above have not been as well accepted. Metadata can be a lot of work! Going back over legacy data and tagging it effectively is very time consuming. It is essential to have a process which is achievable in a realistic time frame.

ANZLIC released its DTD in January 1998 with various updates since then. We shall cover the DTD (NB: see the acknowledgement at end of this chapter) at the time of writing as a way of looking at core spatial metadata concepts.

ANZLIC maintains a comprehensive Web site (http://www.anzlic.org.au/) where readers can find full details of its standards and activities.

9.1.1.1 Constitution of ANZLIC

ANZLIC is a combined initiative of the Australian and New Zealand governments. Australia has six states and two territories, each with largely independent jurisdictions within a federal system. Each state has its own mapping authority and

Surveyor General. The surveyor generals of each state form the core of the ANZLIC governing body. At the time of writing there is no fixed structure across all states and the office of surveyor general may disappear in some places.

9.1.2 The ANZLIC DTD

The top level (zero) element is ANZMETA. Table 9.1 shows elements immediately below. Some of these themselves have sub-elements to a maximum depth of 6. Most of the elements occur once only in a particular order. At the top level, distribution and supplementary information are optional, while the contact information may be repeated. It is fairly straightforward to mark up a dataset according to such a DTD. In the ANZLIC case, there is a complication in that some of the text entries are specified in specific thesauri – a constraint beyond that describable in SGML. We will first run through the basic DTD and return to look at the extension hooks which have been provided for local content and further development.

We shall take an example of real metadata, courtesy of ANZLIC to illustrate the various elements of the DTD. First we start with the **anzmeta** root element. The **anzmeta** element has 9 children in specified order with two elements optional and one which may be repeated and are described in Table 9.1.

Table 9.1 Level 1 Elements

Element	*Content*	*Required*
`citeinfo`	Dataset	Yes
`descript`	Description	Yes
`timeperd`	Data currency	Yes
`status`	Dataset status	Yes
`distinfo`	Access	Optional
`dataqual`	Data quality	Yes
`cntinfo`	Contact information	Repeatable
`supplinf`	Additional metadata	Optional

9.1.2.1 The citeinfo element

There are three required components to the **citeinfo** element: the unique id, **uniqueid**, the **title** and the **origin**, and a **title** element. Of these, only the **origin** element has sub-elements. These sub-elements are: **custodian** and **jurisdiction**. The latter consists of one or more occurrences of **keyword**, which are taken from a prescribed keyword set as discussed in Section 8.4.

```
<citeinfo>
    <title>
        Vegetation: Pre-European Settlement (1788)
    </title>
    <origin>
        <custodian>
```

```
                    Australian Surveying and Land Information
                    Group (AUSLIG)
                </custodian>
                <jurisdic>
                    <keyword>Australia</keyword>
                </jurisdic>
            </origin>
        </citeinfo>
```

This dataset describes a model of vegetation pre-European settlement. Note that some elements have content which is only other elements, giving us a cascade of opening and closing tags.

9.1.2.2 The descript element

The descript element is somewhat more complex with several children. It consists of **abstract, theme** (one or more occurrences and **spdom** (optional) elements in order. Plain text suffices for the abstract. The theme is made up of a series of one or more keywords, which are intended to help the non-expert. The spatial domain, **spdom**, is more complicated. It consists of one or more place elements or a keyword, followed by the bounding element. There may be a neat way of describing the place, say a catchment area or some other legally precise mapping term. In such a case a keyword suffices. In other situations, it is necessary to spell out in geographical coordinates (longitude and latitude) the vertices of each polygonal component (the **dsgpoly** element). Compass coordinates serve to define bounding (**northbc, southbc, eastbc, westbc**).

```
<descript>
    <abstract>
            Shows a reconstruction of natural vegetation of
            Australia as it probably would have been in the
            1780s. Areas over 30000 hectares are shown plus
            small areas of significant vegetation such as
            rainforest. Attribute information includes growth
            form of the tallest and lower stratum, foliage
            cover of tallest stratum and dominant floristic
            type.
    </abstract>
    <theme>
        <keyword>
                FLORA Native Mapping
                VEGETATION Mapping
        </keyword>
    </theme>
    <spdom>
        <place>
        <keyword>
                Australia
        </keyword>
    </spdom>
</descript>
```

9.1.2.3 Data currency: the timeperd element

Data currency is an essential metadata element. One of the big difficulties of effective spatial data mining is the accuracy of the data itself. There is nothing unexpected here. **begdate** and **enddate** elements each have date or keyword sub-elements. Since the concern with metadata has come long after many datasets were created, sometimes the content of the field may be unknown: so instead of a date we may simply have the phrase (keyword) not known!

```
<timeperd>
<begdate>01JAN1780</begdate>
<enddate>Not Known</enddate>
</timeperd>
```

Dates are given in ISO8601 format at present. Note that there is a move within the online community against using text for months. Purely numeric formats are preferred, viz. 2001-04-01.

9.1.2.4 Dataset status: the <status> element

The **status** element records vital information about the development of the dataset. Its current status is the top level element with two sub-elements, **progress** and **update**, which are self-explanatory. The sub-element **progress** is made up of keywords, which are taken from the default thesaurus (complete, in progress, planned, not known). The sub-element **update** has a similar range of update keywords, most of which are self explanatory, e.g. daily, weekly, etc.

```
<status>
     <progress><keyword>Complete</keyword></progress>
     <update><keyword>Not Known</keyword></update>
</status>
```

9.1.2.5 Access: the distinfo element

Three sub-elements describe **distinfo**: **native** describes the stored data format; **avlform** describes the formats in which the data is available (optional, since the data may in fact not be readily available); **accconst** describes any access constraints. Formats may be **nondig** or **digform**.

```
<distinfo>
     <native>
          <digform>
               <formname>ARC/INFO</formname>
               <formname>Vector Data</formname>
               <formname>GINA</formname>
          </digform>
```

```
    <nondig>
        <formname>Maps</formname>
    </nondig>
</native>
<avlform>
<digital>
<formname>Database</formname>
<formname>ARC/INFO</formname>
<formname>Vector Data</formname>
<nondig>
<formname>Maps</formname>
</nondig>
</native>
</avlform>
<accconst>
```

The data are subject to Commonwealth of Australia Copyright. A licence agreement is required and a licence fee is also applicable.

```
</accconst>
</distinfo>
```

9.1.2.6 Data quality: the dataqual element

Data quality is one element where the results are likely to be less than edifying. Much legacy data has been lost, or never had suitable quality indicators. The first sub-element is **lineage**, describing where the data came from; **posacc** and **attrac** give the positional and attribute accuracy; the final elements **logic** and **complete** give the logical consistency and completeness.

```
<dataqual>
<lineage>
Captured from mapping material used to produce
        AUSLIG's 1:5 million
        scale Australia Natural Vegetation, 1989.
</lineage>
<posacc>Not Documented</posacc>
<attracc>Not Documented</attracc>
<logic>Not Documented</logic>
<complete>Australia</complete>
</dataqual>
```

Here we start to see some of the difficulties of handling legacy data: a lot of the characteristics are just not known, were never recorded or have been lost.

9.1.2.7 Contact information: the cntinfo element

The element cntinfo is very straightforward with **cntorg** (contact organisation), **cntpos** (contact position), **address** (mail address(es)), **city**, **state**, **country**, **postal** (postcode), **cntvoice** (telephone), **cntfax** (fax) and **cntemail**.

```
<cntinfo>
    <cntorg>
    Australian Surveying and Land Information Group (AUSLIG)
    </cntorg>
    <cntpos>
        Enquiries to Data Sales Staff, Data Sales,
        National Data Centre
    </cntpos>
    <address>
        PO Box 2
    </address>
    <city>BELCONNEN</city>
    <state>ACT </state>
    <country>Australia</country>
    <postal>2616</postal>
    <cntvoice>
        Australia Fixed Network number
            +61 2 6201 4340
        Australia Freecall
            1800 800 173
    </cntvoice>
    <cntfax>
        Australia Fixed Network number
            +61 2 6201 4381
    </cntfax>
    <cntemail>
        datasales@auslig.gov.au
        mapsales@auslig.gov.au
    </cntemail>
    </cntinfo>
```

9.1.2.8 Metadata: the metainfo element

Now we have the **metainfo** element, which indicates the date the metadata itself was created.

9.1.2.9 Additional metadata: the supplinf element

The final element, **supplinf**, is used to include any supplementary information that is not included above.

```
    <supplinf>
    The Australian Spatial Data Directory(ASDD)
    Also, further information about Spatial Metadata is at
```

```
ANZLIC http://www.anzlic.org.au/metaelem.htm
------------------------------------------------
growth form of tallest and lowest stratum
foliage cover
dominant floristic type
------------------------------------------------
1: 5 million
------------------------------------------------
RESTRICTIONS ON USE
None
------------------------------------------------
3 to 4 mb depending on format
------------------------------------------------
PRICE and ACCESS
RRP $500
</supplinf>
</anzmeta>
```

9.1.3 The keyword element

The keyword element occurs within several elements. It has a required attribute of thesaurus, for which there is a defined list of values. This system works well in restricting keyword use, but does not do quite the full job. Any keyword can take any thesaurus, whereas in fact only a limited range of thesauri are available in each case. The way around this would be a slightly more complex DTD with several different keyword attributes.

9.1.3.1 Comments

The ANZLIC DTD is easy to understand and covers core data elements well. The element names are sometimes cryptic, like the computer programs of yesteryear in which short names were enforced by memory and other limitations. The SGML concrete reference syntax specifies a maximum of eight characters for an element name, but this can be overridden easily and is not necessary in XML. However, with element names largely aligned with US standards, there is not likely to be any immediate change. The use of thesauri is very effective, and the metadata processing tools help to check that the correct data elements are added at each stage.

An alternative to specially developed checking tools would have been a detailed XML schema[1], highlighting an important issue: much of the ground work on spatial metadata has run alongside the development of new recommendations in metadata by the W3C. As a consequence, some of the tools are different or incompatible with emerging web standards, which may impede the development of online GIS.

A related issue is the way metadata documents may be clustered together. At present there is no specification, leaving the organisation to the structure of the

---

[1] We illustrated this sort of approach in Chapter 5.

web site and Spatial Data Directory itself. A more comprehensive object-oriented model, such as implied within RDF, would have the advantages of:

❑   easier searching through a hierarchy;
❑   reduced duplication of data (e.g. where a group of datasets all belong to the same organisation).

   Finally, we have covered above the elements as they pertain to spatial metadata. The DTD also borrows a range of elements from the HTML DTD for describing text structure, such as list elements and paragraphs.

9.1.4 Notes on the ANZLIC framework

Analogous to the use of namespaces in XML, ANZLIC provide a series of thesauri to define terms for the various fields in the metadata DTD. Full details may be found on their Web site (http://www.anzlic.org.au/).
   Marking up documents by hand is time consuming and requires a fair level of skill. In fact adding metadata is a non-trivial task for any organisation. ANZLIC, like other organisations, has tools for entering metadata in a straightforward way. Although the move was made from SGML to XML, allowing new elements to be added at will, there is no explicit provision for extensibility.

9.1.5 Minerals data

Here is a full example of data describing mineral resources. Note that a great deal of information about AUSLIG is repeated.

```
<anzmeta>
    <citeinfo>
        <title>
            Minerals
        </title>
        <custodian>
            Australian Surveying and Land Information Group
            (AUSLIG)
        </custodian>
        <jurisdic>
            Australia
        </jurisdic>
    </citeinfo>
    <descript>
    <abstract>
        Shows the point location of mineral deposits, mines
        and treatment plants in Australia. Attribute
        information includes mine name, State, mine size,
        minerals and status.
```

```
    </abstract>
    <theme>
        <keyword>MINERALS</keyword>
    </theme>
    <spdom>
        <place>
            <keyword>Australia</keyword>
        </place>
    </spdom>
    <timeperd>
        <begdate>Not Known</begdate>
        <enddate>01DEC1990</enddate>
    </timeperd>
    <status>
        <progess>Complete</progess>
        <update>Not Known</update>
    </status>
    <distinfo>
        <native>
            <digform>
                <formname>ARC/INFO</formname>
                <formname>GINA</formname>
            </digform>
        </native>
        <avlform>
            <digital>
                <formname>Database</formname>
                <formname>ARC/INFO</formname>
            </digital>
        </avlform>
        <accconst>
            The data are subject to Commonwealth of
            Australia Copyright. A licence agreement is
            required and a licence fee is also applicable.
        </accconst>
    </distinfo>

    <dataqual>
        <lineage>
            Data for the Minerals database have been gathered
            from a variety of sources including:
            <ul>
                <li>AUSLIG's 1:100 000 and 1:250 000 scale
                topographic mapping material;
                <li>Various larger scale specialist maps and
                plans;
                <li>Mining companies' and State authorities'
                publications, including annual reports and
                maps;
                <li>Numerous private industry publications
                including journals and newspapers; and
                <li>Direct contact with mining companies.
            </ul>
```

```
    </lineage>
    <posacc>
        The horizontal accuracy is fully dependent on the
        source material. At worst the calculated value of
        the feature location is given to the nearest
        minute (approx 1800 metres).
    </posacc>
    <attracc>
        For a given feature code, all attributes listed as
        mandatory are populated. Entries in other fields
        depend on the information availability. The data
        represent the best available at the time of entry.
    </attracc>
    <logic>
        Tests carried out include: check of valid feature
        codes; removal of invalid and system feature
        codes; check for all point features attached to
        the attribute table; tally of all points and
        attribute table records; check of layer/network
        assignment of all features; and cross check for
        invalid feature code to feature type combinations.
    </logic>
    <complete>
        The data were checked through systematic
        comparison against relevant source material.
    </complete>
</dataqual>

<cntinfo>
    <cntorg>
        Australian Surveying and Land Information Group
        (AUSLIG)
    </cntorg>
    <cntpos>
        Enquiries to Data Sales Staff, Data Sales,
        National Data Centre
    </cntpos>
    <address>
        PO Box 2
    </address>
    <city>BELCONNEN</city>
    <state>ACT </state>
    <country>Australia</country>
    <postal>2616</postal>
    <cntvoice>
        Australia Fixed Network number
        +61 2 6201 4340
        Australia Freecall
        1800 800 173
    </cntvoice>
    <cntfax>
        Australia Fixed Network number
        +61 2 6201 4381
```

```
        </cntfax>
        <cntemail>datasales@auslig.gov.au</cntemail>
    </cntinfo>
    <metainfo>
        <metd><date>25NOV1996</date></metd>
    </metainfo>
    <supplinf>
        The Australian Spatial Data Directory(ASDD) Also,
        further information about Spatial Metadata is at
        ANZLIC http://www.anzlic.org.au/metaelem.htm
        ATTRIBUTES
            mine name
            state
            mine size
            minerals
            status
        SCALE/RESOLUTION
            1:1 000 000
        RESTRICTIONS ON USE
            None
        SIZE OF DATASETS
            0.2 to 1.6 mb depending on format
        PRICE and ACCESS
            RRP $300
    </supplinf>
</anzmeta>
```

9.2 METADATA IN THE USA

9.2.1 Overview

North America has of course played a major international role in developing mapping data of all sorts. With a major technological lead in satellites, aeronautics and home of the world's largest imaging companies such as Kodak, Xerox, Bausch and Lomb, the world's computer giants, and many others, this is hardly surprising. The flip side is the vast amount of legacy data.

As we shall see, the metadata movement in the USA has had to deal with issues of *comprehensiveness* and *variations across the states*. The result is a complex standard which is proving difficult to implement in its entirety. As we shall see in this chapter, variations exist at Federal and State level. Moreover, in the attempt to develop a comprehensive federal standard, a complex and perhaps cumbersome standard has emerged.

The USA has been particularly proactive in getting data users, by making data accessible at low cost. Thus the potential for widespread online use is there. But the production of the metadata is slow and hence the ready use of data online may be impeded. First we look at the government bodies involved in data management and some history to metadata development. We then look at aspects of the US metadata standards.

9.2.2 Government bodies managing spatial

The Federal body charged with metadata definition and implementation is the *Federal Geospatial Data Committee (FGDC, http://www.fgdc.gov)*. It received presidential endorsement as Circular A-16 in October 1990. It's only ten years old, so it's easy to see why there should be a huge backlog of data! The FGDC has twelve major sub-committees and six working groups. At the end of 1999 it had endorsed 14 standards with a dozen or more close to final draft.

The next important administrative step came in April 1994, when the *National Geospatial Data Clearing House*, NGDC, *National Spatial Data Infrastructure*, NSDI, was created by executive order 12906 to oversee the development of a *National Spatial Data Infrastructure*, NSDI. The challenge for these organisations was significant. In May 1994 the NSDI began a standards project for a *standard grid reference system*, i.e. there were problems not only on the formats of spatial data but in the very coordinate systems in which it was defined. The Universal Grid Reference Systems was released on May 24, 1999, comparatively recently.

In addition to establishing the NSDI Clearing House, the FGDC was charged with overseeing the development of standards and with encouraging standards implementation, including the provision of grants to non-Federal groups.

Metadata is one of four geospatial standards of the FGDC. Overseeing all such standards is the *Standards Working Group*, SWG, with a complex *modus operandi*. SWG reviews proposals and checks standards for adherence to federal policies. It also liases with external organisations such as ANSI (information technology standards) and the ISO (International Standards Organisation). The principal metadata standard is the *Content Standard for Digital Geospatial Metadata*. It is mandatory for Federal agencies and recommended for state and local governments. Where federal funds are used it may also be required.

9.2.3 Overview of CSDGM

At the time of writing, the CSDGM has 11 categories of data containing several hundred entries. Table 9.2 shows the information categories and the number of elements in each, totalling 245.

Not all entries are mandatory and there is often considerable overlap across categories. There are also two user defined items. Creating such metadata is a skilled job. It requires not only a full understanding of the data itself, but also a detailed understanding of the content standard itself. It involves considerable personnel costs in creating and updating the information.

Table 9.2. Categories of Data for the CSDGM.

Category	Number of elements
Identification	44
Data quality	33
Spatial data organisation	14
Spatial reference	25
Entity and attribute	32
Distribution	43
Metadata reference	13
Citation	15
Time period	8
Contact information	18

9.2.4 Example of New York State metadata

The Adirondacks form a huge park in New York State including the Adirondack mountains and home of the Lake Placid Winter Olympics. Although it is a state park it receives federal funds from the Environmental Protection Agency. As such it is required to use the content standard.

One of the data sets for which metadata has been created is the Watershed Data Layer. The metadata has 511 elements, some of which are duplicates of one another, with one entry repeated no fewer than 14 times! The shortest entries range in size from a single digit to 78 lines of ArcInfo macro language. Some entries are unintelligible to the layman, being nothing more than strings of non-intuitive acronyms, such as the entry *CLIP QDWSCLN QD QDWSCLN NET.* The cost to create this metadata was $10,000, or about $20 per entry.

9.2.5 New York State metadata

Complexity and cost have created a movement towards a simplified standard *MetaLite* which is essentially the mandatory items of the CSDGM, which began at State level and is now taken up by local governments. Metalite has just 29 main elements, specifying data by name and place and providing contact details.

In New York State, the GIS Coordinating Committee has a Clearing House operated by the State Library. It has also a State Data Sharing Cooperative in which all state agencies must participate. Metalite is used and local governments are encouraged to participate also. At the time of writing, late 1999, typical examples of local government involvement included:

❑ Nassau County is using the full CSDGM as a result of Federal Agency support. Their GIS budget is $15 million.
❑ Westchester County may implement MetaLite. Although they have technical staff they are under-funded for any serious metadata activity.

❑ The town of Amherst is ignoring both Federal and State metadata programmes.

The general pattern seems to be that local government finds CSDGM too complicated, too costly and has difficulty in seeing its usefulness.

9.2.6 What is the metadata for?

As we have already discussed metadata serves two functions. In the first phase we simple have the documentation of data. Users will peruse this directly and make arrangements to access or purchase the data. The data will be delivered via a physical transport medium such as magnetic tape or compact disc.

In the second phase, data is available online, and now the metadata can move one step forward to being read by software agents. At this point, it becomes much more important that the metadata is accurate and reasonably complete. Whereas a human operator can wade through data specifications which have all sorts of caveats and comments, this sort of information is not readily processed online by natural language intelligent agents (cf. Section 11.2).

The third phase of development, which we expect to grow rapidly in the next year or so, is that of online spatial queries, where users pay for search results or analysis rather than the data per se. We return to this exciting prospect in the last chapter.

9.3 THE SITUATION IN EUROPE

The UK and Europe have had highly sophisticated mapping technologies and national organisations for a long time. But the European Union would obviously like integration across all the member countries. With numerous different languages, however, this is far from easy. At the time of writing there are many different committees and other bodies, but the many initiatives are far from complete. In this section we attempt to pick out the most important activities and the general direction. Readers are advised to check the book's website for links back to the original standards to keep pace with developments.

Broadly speaking there are three standards under consideration:

❑ The US FGDC which we have considered already in this chapter;
❑ the ISO TC211 international standard, also considered above, and possibly the most important of the three, which should reach its final form in late 2000;
❑ specifically European standards.

The European standards fit into a broader IT initiative, INFO2000, and come in two different levels, CEN and ENV, representing different levels of development. CEN, *Comité Européan de Normalisation,* are essentially the mature standards. But recognising the incredible speed of technological progress the need for fast-track standards led to the ENV, *Euro-Norm Voluntaire,* specifications.

In parallel to these standard mechanisms, there are various organisations linking together the National Mapping Agencies (NMAs) of the member countries. At the top of these is CERCO (*Comité Européan des Responsables de la Cartographie Officielle*)[2], a group of NMAs represented by the Head of each and including a 31 member organisations. In 1993, CERCO spun off the daughter organisation MEGRIN (*Multipurpose European Ground Related Information Network*)[3] but in January 2001, the two organisations merged to form *Eurographics*.

MEGRIN was a non-profit organisation with the status in French law of a GIE (*Group d'Interêt Economique*) with a number of key roles, two of which were:

1. to assist the trade of spatial data across national boundaries and to ensure adequate metadata (such as the GDDD);

2. to harmonise data across national boundaries (such as SABE).

ACKNOWLEDGEMENTS

The USA section of this chapter was based on a presentation given by Professor Hugh Calkins of the University of Buffalo at Charles Sturt University in September 1999 and his help in preparing this chapter is gratefully acknowledged.

The Metadata examples of ANZLIC standards provided in Section 9.1 is Copyright © Commonwealth of Australia, AUSLIG, Australia's national mapping agency. All rights reserved. Reproduced by permission of the General Manager, Australian Surveying and Land Information Group, Department of Industry, Science and Resources, Canberra, ACT. Apart from any use as permitted under the Copyright Act 1968, no part may be reproduced by any process without prior written permission from AUSLIG. Requests and queries concerning reproduction and rights should be addressed to the Manager, Australian Surveying and Land Information Group, Department of Industry, Science and Resources, PO Box 2, Belconnen, ACT, 2616, or by email to copyright@auslig.gov.au

---

[2] http://www.eurographics.org/cerco

[3] http://www.eurographics.org/megrin

CHAPTER 10

Data warehouses

10.1 WHAT IS A DATA WAREHOUSE?

William Inmon (1995) introduced the term data warehousing to describe a database system that was designed and built specifically to support the decision making process of an organisation. However, data warehousing goes well beyond the construction of a database. Data warehousing is a process, *not* a product. The process includes assembling and managing data from various sources for the purpose of gaining a single detailed view of part or all of an organisation.

A data warehouse is an organised collection of databases and processes for information retrieval, interpretation and display. Inmon (1995) defined a data warehouse as a managed database in which the data is:

❑   *Subject-oriented*
    There is a shift from application-oriented data (i.e. data designed to support processing) to decision-support data (i.e. data designed to aid in the decision making process). For example, sales data for a given application contains specific sales, product, and customer information. In contrast, sales data for decision support contains a historical record of sales over specific time intervals.
❑   *Integrated*
    Data from various sources are combined to produce a global, subject oriented view of the topic of interest (Fig. 10.1).
❑   *Time-variant*
    Operations data is valid only at the moment of capture. Within seconds that data may no longer be valid in its description of current operations.
❑   *Non-volatile*
    New data is always appended to the data base rather than replacing existing data. The database continually absorbs new data, integrating it with the previous data (Inmon 1995).

How does a data warehouses differ from a database? Perhaps the most important criterion is that a data warehouse usually contains several distinct databases. The warehouse is an umbrella that links together many different data resources. Now a single database many contain many different tables, but they are tightly integrated within a single software shell. In contrast, a data warehouse is usually created by combining different databases that already exist. These databases may use different software. They may be developed and maintained by completely separate organisations.

Another important criterion is inherent in the subject-oriented nature of data warehouses. Whereas a database normally supports only simple queries, a data

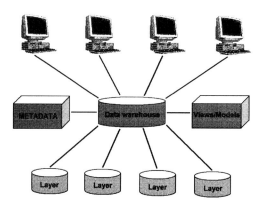

Figure 10.1. The integrated view of geographic data warehouse. Layers from various sources are combined to produce a global subject oriented view of a geographic region.

warehouse usually provides various tools to assist in interpreting and displaying data. This is a consequence of the motivation behind the warehouse. One only takes the trouble to link different databases, if there is useful information to be gained in doing so. So it is always a high priority in a data warehouse to be able to generate the kinds of information for which it was created. For example, many data warehouses are created as commercial marketing tools. A corporation may wish to improve its marketing strategy by analysing sales figures, for instance. So it is imperative to provide tools to extract and analyse the required data.

The final difference between a database and a data warehouse is the scale of the system (Table 10.1). The whole notion of data warehousing arises from the rapidly growing volumes of raw data that are now available in many spheres of professional activity.

Table 10.1. Differences between databases and data warehouses.

Function	Database	Data Warehouse
Size	Mbytes-Gbytes	Gbytes-Tbytes
Nature of queries	Small transactions	Complex queries
Operations	Current snapshot	Historical perspective
Type of data	Raw	Integrated, summarized
Main users	Clerical, operations staff	Analysts, decision makers

Some definitions

As with any new field of endeavour, data warehousing has developed it own set of jargon. Below is a collection of terms used in the remainder of this chapter:

- **Data Mart** or **Local Data Warehouse** is a database that has the same characteristics as a data warehouse, but is usually smaller and is focused on the data for one division or workgroup within an enterprise.
- **Data Transformation**. The modification of data as it is moved into the data warehouse. These modifications can include: data cleansing, normalisation, transforming data types, encoding values, summarising data into time periods, and summarising data in other ways.
- **Data Mining**. The non-trivial process of extracting previously unknown information or patterns from large data sets. A typical application of data mining is to determine what attributes (and values for these attributes) best describe a geographic region. This topic is discussed in more detail later.
- **Knowledge Discovery in Databases** (KDD). A term quite often interchangeably used with data mining. However, KDD concentrates on the discovery of useable knowledge (quite often in the form of rules), where data mining also includes the detection of trends, and model.
- **Z39.50**. ANSI/NISO Z39.50 is the American National Standard Information Retrieval Applications Service Definition Protocol Specification for Open System Interconnections. Z39.50 defines a standard way for two computers to communicate for the purpose of information retrieval. This protocol makes it easier for client applications to connect/query and retrieve information from large databases systems (NISO 1999).

10.2 GEOGRAPHIC DATA WAREHOUSES

The idea of a data warehouse will be familiar to anyone who works with geographic information. Many geographic information systems are essentially data warehouses. This is because a GIS reduces different data sources to the common themes of geographic location and spatial attributes. It is common for different data layers to be derived initially from separate databases. For instance, towns data could be drawn from a database of municipal information, and be overlaid on data drawn from databases on environmental features, health or census records.

Conversely, although they may not be developed with GIS specifically in mind, many data warehouses do have a geographic dimension to them. This is certainly true of most data warehouses that deal with environment and natural resources.

The World Data Center for Paleoclimatology, for instance, acts as a repository for many kinds of environmental data sets that relate to climate change (http://www.ngdc.noaa.gov/paleo/data.html). Data sets such as tree ring records, pollen profiles and ice cores all refer to particular geographic locations.

Many organisations are now setting up geographic data warehouses. For example, the Canadian Government has established a Data Warehouse Infrastructure Project as a project under its National Forest Information Service.

"The objective of the National Forest Information System is to provide a national monitoring, integrating and reporting system for Canada's forest information and changes over time. The current focus of the NFIS Data Warehouse is to provide information to meet Kyoto reporting requirements for forest carbon stocks and criteria and indicators. Over time, NFIS will improve access to and use of accurate and timely spatial and non-spatial information on Canada's forest resources. Some of the information required about the forest is not directly observable (critical wildlife habitat, water quality, risks of disturbance, etc.). The NFIS will integrate the information in the database above with a national modeling framework."

Source: http://nfis.cfs.nrcan.gc.ca/warehouse/

10.3 THE STRUCTURE OF A DATA WAREHOUSE

Data warehousing systems are most effective when data can be combined from many operations system. The nature of the underlying systems, the level of integration required and the organisation structure should drive the selection of a data warehouse structure or topology. Here we present three data warehouse topologies: Enterprise Wide Data Warehouse, Independent Data Mart, and Dependent Data Mart.

The enterprise data warehouse integrates all the information contained in department/working group databases into a single global data repository.

The independent data mart structure implements a number of smaller data warehouses or data marts (Figure 10.2). This structure integrates several databases to produce an independent data mart. This structure is best where organisations have a number of departments with different data needs. One problem with this structure is that there is data replication.

The final structure discussed here is the Dependent Data Mart. Under this scheme, all the underlying databases/applications feed into an enterprise data warehouse (Figure 10.2). A subset of the data warehouse is then inserted into a data mart. This type of structure is appropriate for arrangements where an organisation has a special working group that is interested only in a particular aspect of the organisation entire operations.

The structures for data warehouses described here are not exhaustive by any means. As you could imagine there are many hybrid systems. The overall arrangement of a data warehouse is very much dependent on the organisation and operations of its working groups and decision-making process.

With the exception of Figure 3, all of the data warehouse structures use/rely on data marts. The major question is should data be moved to/from the data warehouse? The movement of data to and from data warehouses or data marts is called data migration. There are two main approaches to data migration.

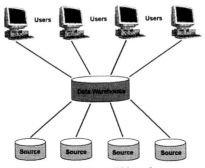

Enterprise Data Warehouse

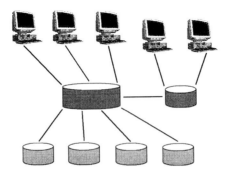

Dependent Data Mart

Independent Data Mart

Figure 10.2.Three common architectures for distributed data warehouses. After Gardner 1996.

The first migration method is the big bang approach. Under this raceme all data is moved in one simultaneous operation. The major benefit of this approach is its speed of data conversion.

The second data migration method is the iterative approach. Under this data migration approach data is moved from/to a data mart/warehouse one system at a time. So warehouses across the organisations are incrementally updated. This approach has less risk of damage. If something goes wrong, only a minimal number of warehouses are damaged.

10.3.1 Indexing of data resources

Closely related to the above issue of the structure of a data warehouse is the question of how the data resources are structured.

Data warehouses operate best where the subject matter can be well defined and where the operating model can be tightly defined. However, the coupling of sites can vary enormously. Databases typically consist of three main components: an interface, data, and tools for search and retrieval. In a distributed database, information and indexes are shared across computers on each of the participating sites. In general, indexed data online can be organised in several ways (Fig. 10.3):

(a) *Centralised* – the traditional model for database access: the entire database resides on a single server; other sites point to it. This is the most common form of network database.
(b) *Distributed data, separate indices at each site* – The database consists of several component databases, each maintained at different sites. A common interface (normally a Web document) provides pointers to the components, which are queried separately.
(c) *Distributed data, multiple queries* – many component databases are queried simultaneously across the network from a single interface. Many search engines adopt this approach.
(d) *Distributed data, single centralised index* – The data consists of many items, which are stored at different sites but accessed via a database of pointers maintained at a single site.

10.4 ISSUES IN BUILDING DATA WAREHOUSES

So you want to build a data warehouse? Where do you start? and what resources will you require? This section summarises some of the major issues, considerations and resources required to develop a data warehouse.

10.4.1 Practical issues

The resources required in developing a data warehouse include the following:

❏ **Cost**. In the development of any type of software project, funding is the biggest issue. Being an enterprise wide information service, the development of data warehouses can be extremely expensive. The key questions to ask here include: How much will the construction of the data warehouse cost? What is the budget for the project? Who will pick up the difference between the budgeted amount and the actual amount? Like other IT projects there needs to be contingencies factored in to account for unforeseen costs (Gardner 1996, Inmon 1996).

❏ **Time**. Time is a critical consideration when developing any software system (Inmon 1996).

❏ **Users**. Any information system is designed for users. If the information system does not deliver the required information to end-users it is practically useless. The needs of the end-users are the most critical factors when developing the warehouse. The warehouse in essences is for the end users not the IT staff! (Hammer et al. 1995)

❏ **People**. Apart from the end users, there is a wide array of additional people that need to have input into the construction of a data warehouse these people include: Support staff, Database Administrators, Programmers, Business Analysts, Data Warehouse Architect, Help Desk staff, Training staff, System Administrators, Data Administrators, Data engineers, Users, Decision-makers, Middle management, Top management (Hammer et al. 1995, Inmon 1996).

❏ **Software and tools**. Data warehousing systems consist of three main components: an interface to the data, tools for searching and retrieval of data (Green 1994). Currently there are literally hundreds of commercial and public domain software packages, data sets, and data analysis tools. The key consideration when selecting software is that the underlying database software is compatible with the selected analysis tools and other database packages. In addition any software package should be extendable to meet the analysis and data requirements in the future. (Worbel et al. 1997).

❏ **Reliability and robustness**.

Just like any software project, the construction of a data warehouse goes through the systems development life cycle (SDLC). The phases that we will discuss here include analysis, design, acquisition, and conversion.

10.4.2 Analysis

The analysis phase is the most important phase in the construction of data warehouse. Failure to correctly analyse the requirements of the system will result in failure. The key issues that need to be addressed are: What data required by the users? What data is the current system(s) collecting? What data is missing? What/where are the source of the data? What are the intended applications of the collected data? What legacy data is stored on legacy media (Gardner 1996)? Is this legacy data useful? Finally one design attribute that should be built in from the start is extensibility. The data warehouse should be extendable in a number of ways. First the warehouse should be able to cope with new data marts being added.

Secondly, it should be possible to add new tables with little effort. Finally, data should be stored in a format that allows new analysis tools to be used to extract information. This leads to the point that the construction of the data warehouse should conform to some form of standard.

10.4.3 Design

The design phase transforms the findings from the analysis phase into a specification that can be implemented. During this phase designers need to decide on a warehouse structure, warehouse schema, level of normalisation, the type of interface. Also the designers need to determine what users should have access to what data, what security measures are to be in-place, and in what mode (online/batch) the warehouse is to operate.

10.4.4 Infrastructure

Being on an enterprise wide scale, chances are that an organisation will not have the required hardware and software infrastructure to support a data warehouse. Hardware requirements include large hard disks, backup and recovery devices, and redundant hardware if the data is mission critical. Fast processors and lots of ram to support many concurrent users and processing of the stored data. Many commercial vendors are producing software products to support data warehousing operations.

The emergence of new standards and protocols may make it easier to create distributed data warehouses in the future. For instance, the *Data Space Transfer Protocol* (DSTP) aims to make it simpler for different databases and systems to share data across the Internet. The idea is that people would convert datasets into a common format, using the *Predictive Model Markup Language* (PMML). Just as the Hypertext Markup Language (HTML) allows people to place documents online in a format that can be universally read and displayed, so the aim of PPML is to achieve the same for datasets. One of the problems encountered in data mining and warehousing is that different people often use different formats to store data.

10.5 ORGANISATION AND OPERATION

How do we organise information on a large scale? In many cases, the sheer scale of a data warehouse demands that the workload be distributed amongst many different organisations. As with any computer based information system, the maintenance of the system is the most expensive activity both in terms of time and cost. This is especially true when the data changes often or when it needs to be updated regularly. In other cases, the warehouse needs to integrate different kinds of data (e.g. weather and plant distributions) that are compiled by separate specialist agencies.

10.5.1 Legacy data

In the early 1970s, virtually all business systems were developed by IBM. These systems were mainframe based, and were implemented with tools such as COBOL, CICS, IMS, and DB2. The 1980's saw mainframes replaced with minicomputers such as the AS/400 and VAX/VMS systems, which run the popular time sharing and client/server operating system UNIX.

In the late 1980's and 1990's desktop computing technology become mature and was rapidly adopted by business. This era also saw an increased popularity in computer networking. Global information networks such as the World Wide Web are still experiencing the growing popularity.

The problem that organisations have encountered in moving from mainframe to minicomputer to desktop computing is the need for data conversion. Many legacy systems still contain valuable information. This need raises several matters.

- **De-normalise data tables**. The normalisation of database tables in common practice. The idea of data normalisation is to conserve data storage space, ensure data integrity, and reduce the chance of data anomalies. However, with data heavily normalised, it may be necessary to de-normalise, to ensure the correct conversion of data into the warehouse.
- **Rules for data integration**. Determine rules for matching data from different sources to allow easy data integration. In some instances simple key matching is not enough.
- **Data cleaning**. When converting data from legacy systems it may be necessary to *clean* the data. This may include activities such as converting data uniform format. For example ensuring that all characters belong to the same character set or carriage returns are consistent. The cleaning of the data may also include the addition, removal or reorder attributes within the legacy data set. Other activities involved in cleaning the data may include: the detection of errors, identify inconsistencies, updating values, patching missing data, and finally the removal of unneeded attributes.

Data conversion is an extremely important step in bringing a data warehouse online. Legacy systems are a valuable source of historic data, however, it may be quite time consuming to convert the data in these old systems. The benefit of including the historical data may outweigh costs involved in converting the legacy data.

10.5.2 Processing objects

One of the most vexing problems with legacy data is to convert it into a form in which it can be used. This conversion can be done permanently by creating new files in the format required. However, permanent conversion may be either impractical or undesirable. For instance, the data may be used in many different ways that would require only parts of the data, each in different forms. So it may be more convenient to extract relevant portions each time they are needed.

Legacy datasets are often stored in idiosyncratic formats and require specific software to extract data from them. One example is data that was created in (say) a commercial format associated with a database program that is no longer in common use. Another case, common for older datasets, is where scientists have stored data in a manner of their own devising, but have provided a program to extract certain elements from the files. If this software is still available, then the most convenient approach might be to keep using the software with the dataset. However, if a warehouse contains numerous legacy datasets, then this approach can become confusing and prone to errors.

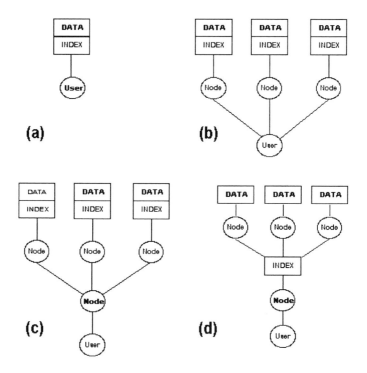

Figure 10.3. Some models for the configuration of online information systems. (a) A traditional centralised database, in which data and index reside at a single site. (b) An unorganised system in which the user must access different sites manually. (c) A common interface to a set of cooperating, but separate databases. (d) An integrated model in which different datasets share a common, centralised index. See text for further explanation.

To impose order on legacy datasets, they can be treated as objects. That is, each dataset, the necessary extraction software, and any scripts needed to automate the extraction are encapsulated as a discrete object (Fig. 10.4). This means that the details of the extraction process are hidden from the rest of the system. As far as users of the data warehouse are concerned, all that exists in the system are objects that can be accessed to supply certain information.

Note that this approach is also an effective method for organising data conversion. Suppose that a processing object (call it SET1) provides methods for outputting data in an interchange format, and that it also has methods for inputting data from the same interchange format. Then that interchange format provides an intermediate stage for converting from SET1 into any other format (SET2 say) that also has the same conversion methods available. This is what makes interchange formats so powerful. A good example is image formats. There are at least 100 different formats in existence; many now rarely used. To provide direct converters between each pair of formats would require 9900 filter programs. However, by passing them through a single universal interchange format, the number of required programs drops to 200, one input and output filter for each format. In practice several interchange formats are needed to take into account differences in the nature of the data, especially raster versus vector graphics.

10.6 DATA MINING

Data mining, as the name suggests, is the process of exploring for patterns and relationships that are buried within data.

Now that you have all the data that you need in a data warehouse what do you

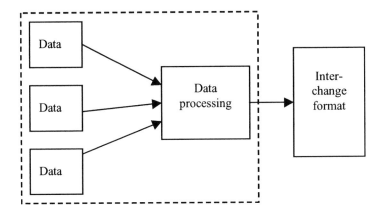

Figure 10.4. Operation of a warehouse object (enclosed by dashed line) to extract legacy data.

do with it? A common process performed on data warehouses is data mining. Data mining or knowledge discovery in databases was first suggested in 1978 at a conference on the analysis of large complex data sets (Fayyad 1996a,b). Although the field of data mining has been around for over two decades, it has been only in the last five years that the required data storage (data warehousing) technologies have emerged.

Data mining is an inter-disciplinary area of research. It draws on technologies from database technology (data warehousing), machine learning, pattern recognition, statistics, visualisation, and high-performance computing (Fig. 10.5). The central goal is to identify information *nuggets*. These are patterns and relationships within data that are potentially revealing and useful.

- ❏ **Machine learning**. The area of machine learning focuses on developing computer systems that can adapt and learn (induce knowledge) from examples or data (Dietterich 1996). The systems aim at moving the traditional view of computer programs from:

  program = algorithm + data

  to the more elaborate

  program = algorithm + data + domain knowledge

  The domain knowledge is acquired from prior experiences solving similar problems (Michalski et al. 1998, Aha 1995).

- ❏ **Statistics**. Many of the machine learning techniques, and other algorithms, centre on detecting the frequency of certain patterns and relationships. Non-parametric statistics can be used to perform hypothesis and exploratory data analysis. A number of references suggest that data mining is a practical application of statistics (e.g. Fayyad et al. 1996b, Eklund et al. 1998, Ester et al. 1999, Chawla et al. 2001).

- ❏ **Pattern recognition**. The pattern detection can take on a number of forms. Pattern recognition can be determining what things commonly appear with each other. Similarly it can be the detection of what attributes are deterministic of other attributes.

- ❏ **High performance computing**. The central idea in data mining is to sift through large volumes of data to detect patterns. Data warehousing requires the storage of large volumes of data. This technology has become available only in the last 3-5 years. Likewise, data mining requires fast access and lots of internal storage (RAM) to efficiently process the data in a data warehouse. With the increase in computer technology in the last couple of years, the technology has become affordable, most current PC, and low-end workstations are now powerful enough to perform data mining and data warehousing functions (Fayyad et al. 1996a,b).

Techniques used in data mining to extract information include: Artificial Neural Networks, Genetic Algorithms (Fayyed et al. 1996), Radial Basis Functions, Curve Fitting, Decision Trees (Quinlan 1993), Rule Induction (Fayyed et al. 1996), and

Nearest Neighbours algorithms. Each of these techniques can be used to extract different forms of information (Kennedy et. al 1998).

10.7 EXAMPLES

10.7.1 The Distributed Archive Archive Center (DAAC)

The Distributed Active Archive Center (DAAC) was established in 1993 and consists of eight data archiving centers (Alaska SAR Facility (ASF), CIESIN-SEDAC, EROS Data Center, NASA/Goddard Space Flight Center, Jet Propulsion Laboratory (JPL), NASA/Langley Research Center (LARC), National Snow and Ice Data Center (NSIDC) and Oak Ridge National Laboratory (ORNL)) and two affiliated data centers (National Oceanic and Atmospheric Admission Satellite Active Archive (NOAA-SAA) and Global Hydrology Research Center (GHRC)) (see URL's [3–12]). The DAAC is part of the Earth Observing System which is an integral part of NASA's Earth Science Enterprise.

Historically, scientists have had difficulty conducting interdisciplinary research because locating useful data required contacting many different data centers for data holdings and availability. EOS designed the DAAC to overcome the problems associated with conducting interdisciplinary earth science research.

DAAC offers over 950 different data sets and products. Including:

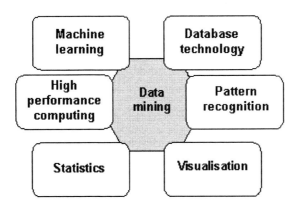

Figure 10.5. Disciplines associated with data mining.

❏   Satellite imagery;
❏   Digital aerial Images,
❏   Airborne sun photometer,
❏   AVHRR –Advanced Very High Resolution Radiometer,
❏   Thermal infrared multispectral scans,
❏   Field sun photometer,
❏   Other data sets.

DAAC allows the easy integration of data from different data sets by ensuring all data/image files are in a common file format, data sets are adequately documented, and contain sufficient metadata.

The warehouse contains data sets collected from a range of studies. Examples include OTTER (Oregon Transect Ecosystem Research) and FIFE (First ISLSCP (International Satellite Land Surface Climatology Project) Field Experiment). In addition there are periodic data sets collected by regular surveys.

The DAAC makes large percentage of the data sets available at no cost. These data sets can be downloaded from the relevant data center via FTP. Some data sets are available on CDrom and can be obtained at a small fee.

An interesting point that DAAC highlights is that the management of such a data repository by one organisation is almost impossible. However, by dividing the work between various organisations and by making the organisations responsible for the management and administration of the data, such a project becomes possible.

10.7.2 Wal★Mart

Wal★Mart [WWW 16] is a chain of over 100 retail stores that sell a wide range of household goods, electrical appliances, toys and other goods. The Wal★ Mart chain also includes, services stations, sports stores and discount retailers, distributed across the USA.

In the late 1980s and early 1990s, Wal★ Mart was having difficulty managing inventory, coordinating operations between the general office, distributors and retailers. In addition, Wal★ Mart's current information systems had a number of limitations. These included:

❏   Limited data accessibility,
❏   Three month detail data on-line (limited),
❏   Detail History on tape for several years (with poor accessibility),
❏   Hundreds of generalised short term paper reports.

The overall environment was described as *"Data Rich but Information Poor"* (Hubber 1997). While Wal ★Mart has large volumes of data, the transformation of data into information was hindered by: (1) the distributed natuer of the databases, (2) limited time view of the operations of the organisation and (3) limited access to the data.

The grand challenge to Wal★ Mart was to implement a system where all the data was located in one central repository. The data could be accessed from any platform; all the data were accessible; the database could support hundreds of users; data accessibility needed to be on a 24 hours a day, 7 days a week basis; and users needed to have timely access to the data.

The solution to the problem was to construct a centralized data warehouse, which had standardized access. The Wal★Mart data warehouse has grown to be one of the largest data warehouse (Hubber 1997) in the world. At the end of 1998 the Wal★Mart data warehouse was 10 TB in size, and it was expanding at a rate of between 200–300 Mb per day. The largest table in the warehouse contains 20 billion rows and can support 650 concurrent users (Wiener 1997).

A number of benefits flowed from the data warehouse. These included:

❑ Better markdown management,
❑ Improved next session planning,
❑ Improved stock control,
❑ Increased leverage during vendor negotiation,
❑ Improved long-term forecasting and trend analysis.

The use of a data warehouse has allowed Wal★Mart to gain a strategic advantage over its competitors, and suppliers.

10.7.3 Geographic data warehouses

At the time of writing, many GIS sites would claim that they operate a data warehouse. However, very few of the geographic data warehouses are truly distributed. In most cases, the "network" really consists of a single coordinating site that receives datasets from a network of participating sources. This situation is partly geographic. In many cases the participants are all agencies that deal with different kinds of information (e.g. labour, health, environment), but cover the same region.

The GIS Centre of King County, Washington, USA, is perhaps typical of this approach. It coordinates a spatial data warehouse that relies on a network of sites.

> "In King County, data and *metadata are received from a distributed network of GIS sites and checked into a common spatial data warehouse organized by subject areas. Each site has data stewardship responsibilites to assure that information is comprehensive and trusted across functional units as an enterprise information base.*
> *"Instead of a project oriented approach that requires spatial data conflation KCGIS uses framework coverages to assure that a variety of spatial data will correctly reference or 'nest' other spatial data."*
> KCGC (2001)

In most cases the King County raw data are sold and distributed as ARC/INFO coverages or ArcView shapefiles on CDs. The Centre also produces maps on demand. The online metadata includes feature attribute tables, coverage descriptions and data overviews, and contraints and disclaimers.

An example of a truly distributed information system is the ANZLIC Australian Spatial Data Directory (http://www.environment.gov.au/net/asdd/). The Metadata Working Group of the Australia–New Zealand Land Information Council has established an online interface that enables users to query the databases of all its members simultaneously. The metadata directory itself is spread across 20 or more separate Web sites. Each query spawns a set of subqueries that are farmed out to each site in the network. At each site, the query runs on the metadata held by that agency. The user can select any desired subset of sites to query. Each site automatically carries out the query of its local database and returns data to the interface for browsing by the user. While this process is under way the interface site displays and updates a summary table of the returns. When the query is complete, the user can browse the results returned by each site.

10.8 STANDARDS FOR ONLINE GIS WAREHOUSES

At the time of writing this account, several standards are in the process of being developed that will substantially simplify the process of creating an online data warehouse, especially for geographic information. Some of these standards are ones that we have described in earlier chapters. However, it is worth touching on them again briefly here.

Some standards are specific to geographic information. Others are generic: they deal with any kinds of information. Mostly these generic standard derive from activities of the World Wide Web Consortium, others are specific to data warehouses, and others to the Open GIS Consortium (see Chapter 7).

10.8.1 The Web Map Server Interface Standard

The specification for the OpenGIS Web Map Server Interface (OGC 2000a), defines a Web Map Server to be a site that can do three things:

1. Produce a map (as a picture, as a series of graphical elements, or as a packaged set of geographic feature data),
2. Answer basic queries about the content of the map, and
3. Tell other programs what maps it can produce and which of those can be queried further.

The importance of this design is that it means that resources from different servers can be combined without the need for direct coordination between the two organisations involved. For instance, it makes it possible to extract (say) a satellite image from one site and overlay it on a map (taken from a second site) of the region covered by the image.

The WMSI model divides data handling into four distinct stages:

1. Filtering data from a data source using query constraints.
2. Generating the display elements from features in data according to the required style.

3. Rendering the image.
4. Display the image in a given format.

One advantage of this hierarchy is that it allows some flexibility in the way data are delivered. For instance, if a client is capable of rendering maps that are encoded using scalable vector graphics, then the map could be delivered in that format, whereas a "thin client", with no rendering capability at all would need a map that had been converted into (say) a GIF or JPEG image.

10.8.2 Geographic Markup Language (GML)

Another important tool for coordinating the delivery of geographic information across many different sites is GML (see Chapter 7). The Open GIS Consortium (OGC 2000b) describe the *Geographic Markup Language* (GML) in the following terms:

> *"The Geography Markup Language (GML) is an XML encoding for the transport and storage of geographic* information, *including both the geometry and properties of geographic features. This specification defines the mechanisms and syntax that GML uses to encode geographic information in XML. It is anticipated that GML will make a significant impact on the ability of organisations to share geographic information with one another, and to enable linked geographic datasets."*

The importance of GML is that it provides a standard for using XML to markup GIS objects.

10.8.3 The Predictive Model Markup Language (PMML)

The Data Mining Group (DMG) describes itself as "a consortium of industry and academics formed to facilitate the creation of useful standards for the data mining community". The group's online service is currently hosted by the National Center for Data Mining at The University of Illinois at Chicago. Perhaps the group's chief contribution to date has been the development of the Predictive Model Markup Language (PMML). The Group, which released PMML 1.1 in August 2000, describes PMML as follows (DMG 2000):

> *"The Predictive Model Markup Language (PMML) is an XML-based language which provides a quick and easy way for companies to define predictive models and share models between compliant vendors' applications.*

> *"PMML provides applications with a vendor-independent method of defining models so that proprietary issues and incompatibilities are no longer a barrier to the exchange of models between applications. It allows users to develop models within one vendor's application, and use other vendors'*

applications to visualise, analyse, evaluate or otherwise use the models. Previously, this was virtually impossible, but with PMML, the exchange of models between compliant applications now will be seamless."

PMML provides DTDs for marking up many types of models, including statistics, normalisation, tree classification, polynomial regression, general regression, association rules, neural network, and centre-based and distribution-based clustering (DMG 2000). It also includes methods for specifying data dictionaries and mining schema.

Here, for instance, is an example of PMML code to define a simple data dictionary. In this case the data dictionary defines three kinds of variables:

- a categorical variable called *landcover*, which defines a land classification and can take three possible values "Forest", "Farmland", or "Grassland";
- an ordinal variable called *roadtype*, which assigns ranks to different kinds of roads; and
- a continuous variable called *elevation*, which takes numbers as its values.

```
<data-dictionary>
<categorical name="landcover">
    <category value="Forest" />
    <category value="Farmland" />
    <category value="Grassland" />
    <category value="." missing="true" />
</categorical>
<ordinal name="roadtype">
    <order value="Freeway"  rank="4" />
    <order value="Highway"  rank="3" />
    <order value="A"        rank="2" />
    <order value="B"        rank="1" />
    <order value=" Track "  rank="0" />
    <order value="."        rank="N/A" missing="true" />
</ordinal>
<continuous name="elevation">
    <compound-predicate bool-op="or">
        <predicate name="elevation" op="le" value="1" />
        <predicate name="elevation" op="ge" value="10" />
    </compound-predicate>
</continuous>
</data-dictionary>
```

The main function of PMML is to define models. As an example, let's take the case of defining a decision tree model using PMML. A decision tree consists of a hierarchy of linked nodes, with a test condition at each node. For instance, a condition such as *"elevation* GT 1000" will return TRUE if the local elevation is greater than 1000 metres, and FALSE if not. These two outcomes (TRUE or FALSE), thus specify two branches in a tree. We can then attach other test conditions further down each branch. At the end of each branch is a so-called *leaf node*. In the decision tree values are specified at each leaf node.

Decision trees are common in classification problems, such as interpreting data in a satellite image. In the following simple example of PMML code (based on details of the language given in DMG 2000), we define a simple decision tree that is based on the values of two variables: *var1* and *var2*. Notice the use of XML format.

```
<?xml version="1.0" ?>
<pmml version="1.0">
<data-dictionary>
    <continuous name="var1" />
    <continuous name="var2" />
</data-dictionary>
<tree-model model-id="classify01">
<node><true/>
    <node>
    <predicate attribute="var1" op="le" value="0.5" />
        <node score="1">
        <predicate attribute="var2" op="le" value="0.5" />
        </node>
        <node score="2">
        <predicate attribute="var2" op="gt" value="0.5" />
        </node>
    </node>
    <node>
    <predicate attribute="var1" op="gt" value="0.5" />
        <node score="3">
        <predicate attribute="var2" op="le" value="0.5" />
        </node>
        <node score="4">
        <predicate attribute="var2" op="gt" value="0.5" />
        </node>
    </node>
</node>
</tree-model>
</pmml>
```

10.8.4 Data Space Transfer Protocol (DSTP)

Closely related to PMML is the *Data Space Transfer Protocol* (DSTP), which is being developed by the National Center for Data Mining, at the University of Illinois at Chicago (NCDM 2000). DSTP is a proposed standard that aims to simplify online data mining activity (NCDM 2000). The goal is to make it simpler for different databases and systems to share data across the Internet. As mentioned earlier in this chapter, the idea is that people would convert datasets into a common format, using the Predictive Model Markup Language (PMML). Just as the Hypertext Markup Language (HTML) allows people to place documents online in a format that can be universally read and displayed, so the aim of PPML is to achieve the same for datasets. One of the problems encountered in data mining and warehousing is that different people often use different formats to store data.

> *"Currently, data storage on all platforms is executed in an ad hoc fashion. Even as new applications are created, new formats for storing data associated with those applications are created. This creates an enormous challenge to other users and applications that wish to access this data, but do not wish to be constrained to a specific platform or application. As HTTP, HTML, web servers and browsers introduced a way to share documents across different platforms, DSTP, DSML, and data servers and clients introduce a platform independent way to share data over a network. DSTP relies only on data storage concepts (currently columns and rows), and is independent of the type of data storage used, whether it be files, database, or a distributed data warehouse structure. DSTP makes it possible in one location to locate, access, and analyze data from several other locations. DSTP also reduces the dependency on the data file, because it correlates data based on common keys in different data sets. DSTP allows the true conceptualisation of a data space."* (NCDM 2000)

10.9 THE FUTURE OF GEOGRAPHIC DATA WAREHOUSES

Several trends are driving the spread of data warehouses:

❑   The first trend is the growth of computer storage capacity and processing speed, which make the data warehouse a practical reality.
❑   Secondly, the increasing supply of data is now a routine, automated by-product of many commercial and professional activities.
❑   Finally, the growth of the Internet makes it both feasible and necessary to organise data collection and dissemination on a large scale.

What might an environmental information warehouse look like to a user in a few years from now? One vision of the future of the Internet is what we call the "Knowledge Web". The present emphasis on sites and home pages will disappear. Instead the user will simply look for information about a topic and be guided to that information by an intelligent system that actually teaches you as you go. This view would also apply to (say) a world environmental information warehouse. Suppose for example that a student wanted to know about conservation of plants and animals in the local area. Starting from some general heading (say plants) the system might guide the student through relevant topics (*e.g.* biodiversity, conservation, geographic information) at each stage providing background information and links to other information. A geographic query might involve selecting an area on a map and choosing what to be shown from a range of choices offered.

Suppose alternatively that a public servant wanted to see a report about (say) natural resources in southwest Tasmania. After selecting the exact area and topic she might use a report generator to select the kinds of items she wanted to include.

The choices might cover a standard list of resources, time period, geographic area, type of material (e.g. policy papers, scientific studies, educational material) and types of items (e.g. maps, tables, graphs, text etc.). The system would then build a preliminary report with the option of going back and exploring any aspect in more depth.

Systems such as the above are not far off. Data warehouses (including distributed data warehouses), and associated technologies such as data mining, already constitute a new paradigm for the collation, interpretation and dissemination of information. In some areas of research, such as molecular biology and astronomy, the growth of large public domain repositories has revolutionised the way scientists go about their work. In other areas, especially environmental management, the development of data warehouses is crucial to the future effectiveness of planning and management.

CHAPTER 11

New technologies for spatial information

In previous chapters, we have looked at elements of the current technology for placing geographic information services on the World Wide Web. Many of these technologies are still in their infancy. At the time of writing, even XML, and as a result all of the standards that hang off it, has not yet come into general use. One reason for this is protracted discussions about the nature of the translation language and processes for linking XML documents to style sheets (XSL). However, once these issues are resolved, we can expect to see rapid implementation of GML, SVG, PMML, DSTP, and all the other standards that we have considered. We anticipate that the implementation of these standards and protocols will lead not only to an explosion of activity in online GIS, but also to entirely new kinds of applications. In this chapter we explore some of the possibilities that we foresee. Some of these are already happening in prototype or experimental form.

11.1 VISIONS OF A GLOBAL GIS

In an era of increasing globalisation, more and more issues demand that managers, planners and policymakers in every sphere of activity must put their decision making into a global context. Business is increasingly international, not only through multi-national corporations, but also through electronic commerce, global stock markets and currency exchange.

The same is true in the areas of health, society, environment, and government. In health, for instance, international travel means that diseases such as AIDS are no longer regional concerns, but worldwide problems. Enhanced communications mean that culture and social values are becoming universal. In this context, it is highly desirable to create an online system that encompasses geographic information about any issue anywhere.

And the need for on-the-spot, up-to-date information is not confined to large organisations. To take an example, let's suppose that a young married couple living in Edmonton Canada look to investments as a way of boosting their savings and securing their future. So they look at prospects not only within Canada, but also around the globe. In the evening they go online and check out the stock exchanges in Australia, Tokyo and Hong Kong. In the morning they do the same for London and Bonn. If they find companies they are interested in, then they naturally want to find out more. So they might be looking at such widely spread prospects as a tour company based in Dunedin, New Zealand; a chain of micro-breweries in Portland, Oregon, or a company building intelligent robots in Edinburgh, Scotland. In each case they would probably want to access detailed local information, not only about the company, but also about the area, local competition, and so forth. In short, they need to be able to access detailed geographic information from all over the world.

We could find similar stories for many other areas of activity as well. They all point to the need for rapid access to geographic information from all over the world. In the following section we look more closely at one particular issue, the problem of global environmental management in more detail.

11.1.1 The problem of global environmental management

One of the great challenges facing mankind at the turn of the millennium is how to manage the world's environment and its natural resources.

The problem is immense. The planet's surface area exceeds 509,000,000 square kilometres. Simply monitoring such vast tracts is a huge task. The total number of species is estimated to be somewhere between 10 million and 100 million. At the current pace it would take at least another 300 years of taxonomic research simply to document them all. Modern technology can help with these tasks, but at the same time generates huge volumes of data that must somehow be stored, collated and interpreted.

The problem is also acute. As human population grows the pressure on resources grows with it. We have now reached a point where virtually no place on earth is untouched by human activity, and where it can be questioned whether the existing resources can sustain such a large mass of people indefinitely. Slowly we are learning to use resources more carefully.

Given the size and urgency of the problem, piecemeal solutions simply will not do. We have to plan and act systematically. Governments, industry and conservation all need sound, comprehensive information from which to plan. The problem is so huge that nothing less than the coordinated efforts of every agency in every country will be adequate.

Our ultimate aim should be nothing less than a global information warehouse documenting the world's resources. Until recently such a goal was unattainable. Collating all available information in one place is simply not possible. However, improvements in communications, and especially the rise of the Internet as a global communications medium, now make it feasible to build such a system as a distributed network of information sources.

11.1.2 Prospects and issues

The logical endpoint to putting geographic information online is to create a comprehensive, global GIS. If information from different sources can be seamlessly combined into a single resource, then there is nothing in principle to prevent such a system. Here we briefly consider what such a system might look like and what would need to be done to put it into practice.

First, it is clear that many relevant services already exist. Most of them provide raw material that could easily become components of a global GIS.

❑ Many online services already provide worldwide map coverages (e.g. Xerox Parc, CSU Mapmaker). Most of these online services are based on the Digital Chart of the World (DMA 1992, Danko 1992). So it is already possible to draw maps, down to 1 km resolution or better, for anywhere in the world. In some cases the basic data are augmented by additional layers.

❑ There are also many sites and services that provide detailed maps or geographic queries for particular countries or regions.

❑ Other sites provide global coverages or queries for particular themes or data layers. These cover geographic layers of many different kinds, such as physical, biological, economic, and political.

❑ Huge numbers of sites provide detailed online information about particular geographic objects, such as towns and parks.

The challenge for a global GIS online is to integrate all of these resources into a single overriding service. This is not a new idea. There are plenty of precedents to work from. Integrated online services, many of them global, already exist. They provide striking proof that a comprehensive global GIS is a practical possibility.

Probably the best examples are services that deal with tourist information. In almost every case, the service provides systematic indexes that link to large networks of online sources of geographic information. Many of these geographic networks are highly specific in nature, such as geographic indexes of hotels and other accommodation. They are sometimes subsidised by a particular industry group. However, some networks have taken a broad brush approach from the start. Prominent examples include the Virtual Tourist, and the Lonely Planet travel guide (Lonely Planet Publications 2000).

Now it could be argued that there is no need to set out to build a global GIS from scratch. Such systems already exist. Some of the services we refer to above are very impressive in the range and depth of information they supply. However, most of the current examples are responding to a particular commercial need and opportunity. There are many kinds of studies for which existing information resources are totally inadequate.

Some concerns are already being covered by international cooperation between governments. For instance, the Global Biodiversity Information Facility (see Section 6.1) is a model framed in the context of international agreements on biodiversity conservation. Again, such networks are responding to a perceived need, this time environmental, rather than commercial.

The above points raise the question of just why is a global GIS needed? One answer is that in an era of increasing globalisation, people need to be able to access and combine many different kinds of detailed information from anywhere on Earth. Also, another aspect of globalisation is that virtually every activity impinges on everything else. So the proponents of a commercial venture need to know (say) about environment, social frameworks, and politics so they can be prepared for possible impacts and repercussions. Likewise, in (say) conservation, managers and planners need to be aware of the potential commercial, political and other consequences of banning development in particular regions.

Given the above needs, the information coverage of a global GIS needs to be comprehensive. It also needs to be scalable. That is, users need to be able to zoom in and obtain regional maps and data coverages at any desired scale.

So what is involved in creating a truly global GIS? Many of the formal technical requirements have been outlined in earlier chapters, especially Chapter 6 on networks, and Chapter 10 on distributed data warehouses. From these and other parts of our account, it will be clear that international agreement on standards and metadata for online GIS is essential. Clearly the first step, which is essential to get beyond large scale mapping, is agreement between national mapping agencies about the online provision of fine scale geographic data, such as topography and cadastral layers. E-commerce models need to be developed for the sale of data online.

The actual mechanics of running a global GIS are flexible. We suggest that an imposed system is not necessarily desirable. To be all inclusive, such a system would have to impose standards and constraints on all manner of information. The complexities involved are likely to be so great that either the system never gets off the ground, or else it would become so large and unmanageable that to do so would be more trouble than it is worth. A better approach is to set up a general framework that helps the main elements of a system to emerge naturally, driven by user demand. The challenge is to provide a framework that both ensures that integration happens, and yet prevents the elements from becoming idiosyncratic.

The development of metadata standards, notably XML, is intended to encourage integration of this kind. However, there is still plenty of scope for divergence of standards. For instance, different industries, or professions could develop incompatible namespaces, data dictionaries and other standards, that make integration impossible. It is these kinds of issues that governments, not to mention every organisation involved in the process, need to pay careful attention to.

However, as we saw in Chapter 7, metadata is set up to describe data content and accuracy. Although it contains ownership details, there are inadequate hooks to make truly flexible e-commerce possible. Examples of additional requirements might include the following:

❏ *purchasing model* – an intelligent agent, or a human browser, may wish to purchase data in large chunks, or merely to answer a single query; it may pay by credit card, existing account or some other smart cash mechanism;
❏ *security model* – data may be available only to particular clients. For example, it might be available only to citizens and approved national allies. To verify that an agent has the required privileges, appropriate authentication methods would be needed;
❏ *copyright model* – related to security, the data might contain watermarks or other devices that render it impossible to copy and suitable only for certain uses (i.e. with approved software).

The object-oriented approaches that we have promoted here form a crucial part of the above process. Though (say) hotel listings and species occurrence lists are almost like chalk and cheese, they do share important elements in common. It is important that those elements (e.g. source, custodian, geographic location) are identified as discrete objects which diverse information resources can use in a consistent way.

What we are talking about above is the tension that always exists in information studies between top-down (i.e. imposed from above) and bottom-up (i.e. emerging from below) approaches to problems. We suggest that the best way to create a global GIS is through a balanced combination of both methods. Governments (and other organisations, such as W3C), need to provide basic top-down guidelines that both permit and encourage individual, bottom-up initiatives, but at the same time ensure that they conform to certain basic principles.

Whilst XML and other features of the Web allow these bottom-up activities to occur, the exact mechanisms require some discussion, which we pick up in the next section.

11.2 INTELLIGENT SYSTEMS

In this book, we have described the use of metadata to organise online sources of information. Although we have stressed many of the issues involved, an important concern remains. That is, how can we use that information most effectively. Knowledge based and intelligent systems provide a natural way of working with both metadata and data content. They also have potentially important applications for geographic information.

11.2.1 Example – Mapquest

A good example of applying problem-solving intelligence to geographic problems in an online geographic information service is provided by the Mapquest route finder (Fig. 11.1). The company Mapquest provides an impressive free service online that makes use of sophisticated problem-solving algorithms. Users of the service can obtain detailed instructions for driving between any two locations in North America. The input form (Fig. 11.1a) asks for two street locations including the city, state and country. The system then works out the best route and returns the results to the user. The output (Fig. 11.1b) includes maps, with the route marked, as well as detailed driving directions.

11.2.2 Knowledge based systems

Doing is knowing. The essence of knowledge is the ability to do things. If information is data that has been distilled, so that essential patterns and relationships are made clear, the knowledge goes one step further. It includes details of how to use information.

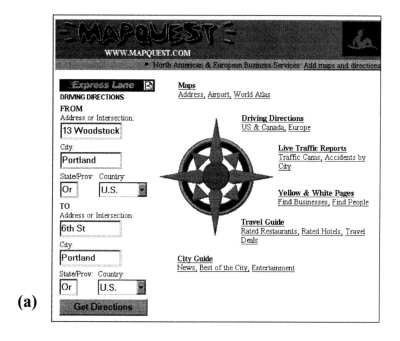

(a)

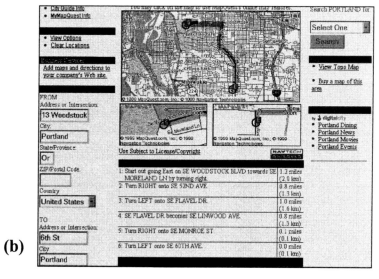

(b)

Figure 11.1. Online route direction finding service from Mapquest. The system gives detailed instructions for driving between any two locations in North America. (a) The query interface. (b) The resulting street map with the route drawn in and instructions provided.

Knowledge is usually expressed in the form of rules. Rules take the general form P→ Q (read "P implies Q" or "If P then Q"), where P and Q are logical statements. The following simple examples illustrate some of the kinds of rules that might be used in an intelligent ecotourism information system.

If X is a park and is near a major city then X has ecotourism potential.

If population of X > 100,000 and X has an airport then X is a major city.

If the distance between X and Y is less than 200 kilometres then X is near Y.

The following set of rules show the sort of query that could be run in trying to locate Web sites that are worth searching for geophysical information.

If site X is relevant to geology then search site X.

If site X is useful and site X is near site Y, then site X is relevant to Y.

If site X is in list of reference sites then site X is useful.

If site X key words contain the word Y then site X is a "Y site".

Many systems incorporate rules, either explicitly or implicitly. Most electronic mail programs allow users to filter incoming mail. For instance, they may place copies of messages in different folders based on words in the subject line, or on the sender's address. Databases and spreadsheets can include rules (like those shown above) that convert data based on values of particular attributes. For instance, the spreadsheet program Excel allows user to create look-up tables by which essentially encapsulate a series of rules detailing how values of one attribute depend on values of another.

11.2.3 Expert systems

Expert systems are computer programs whose main function is to incorporate and use knowledge about a particular domain. For instance, a geographic expert system about ecotourism might include rules to identify sites that have potential for developing visitor programs. A geophysical expert system would contain rules about sites worthy of exploring for particular kinds of minerals. An expert system of web sites would contain rules about how to locate sites that might provide particular kinds of data.

Expert systems often contain hundreds, or even thousands of such rules. The usual search mechanism is a procedure known as backward chaining. The system starts from the statement that it needs to satisfy and works backwards. For any potential solution X it tries to form a chain of clauses $P(X) \rightarrow Q(X)$, $Q(X) \rightarrow R(X)$ etc. in which the precondition $Q(X)$ in one rule is the conclusion in the previous rule. This process continues until the program finds a precondition that it can check directly.

Expert systems are usually produced using a development shell, together with an appropriate logical language. Common languages include LISP and Prolog, however, most shells include their own scripting language. A popular shell and language for developing intelligent systems online is CLIPS (Giarrantano and Riley 1989). This freeware system includes modules for incorporating expert systems into Web services.

The following short sample of CLIPS code asks the user to define a search radius (rule1). It then runs a test to determine whether a point located a certain distance from the search centre lies within a circle of that radius (rule2).

```
(defrule rule1
   =>
   (printout t "What search radius do you want? " crlf)
   (assert (radius = (read))))

(defrule rule4
   (radius ?r)
   (distance ?d)
   (test (< ?r ?d))
   =>
   (assert (accept_point yes)))
```

11.2.4 Adaptive agents

In building an expert system, the expert tries to incorporate all the knowledge that is necessary to solve a particular kind of problem before the system is used. This may involve a long development procedure. However, there are many kinds of problems for which complete knowledge is unobtainable. In such cases, acquiring new rules and data need to be an on-going activity of the system. For example, in searching for information on the Web, new sites are always appearing and new data is constantly being added. So systems must be able to constantly add details of such items to its information base.

Adaptive agents are software programs that are intended to cope with this kind of problem. The term agent has many meanings that are relevant here. One sense covers programs that act on behalf of a user. For example, most Web search engines use software agents that automatically trawl through web sites, recording and indexing the contents. Another, related definition concerns programs that automatically carry out some task or function. The term adaptive refers to software that changes its behaviour in response to its "experience", that is, the problems that it works on. In general the software agents used by search engines are not adaptive. Although they accumulate virtual mountains of data, the way they function is essentially unchanged.

One way for an agent to adapt is by adding to its store of knowledge. That is, it adds new rules to its behavioural repertoire. This idea is perhaps best illustrated via an example. Suppose that to carry out a particular kind of Web search, the user feeds the agent with a set of rules that provide explicit instructions. Under these conditions, not only could the agent carry out the query, but also it could add the instructions to its knowledge base. This way, future users would be able to carry out the same query without needing to instruct the agent how to go about it. Of course, to ensure that future users know that such a query is available, it needs to have a name by which it can be invoked.

However, agents can go further in the learning process. Let's suppose that we set the agent to work to provide a report about (say) kauri trees in New Zealand. To carry out the query we give the agent a set of appropriate rules, and we supply a name, NZKAURI say, for the query. Then this query becomes a new routine that future users can recall at any time simply by quoting its name. However, it is limited by its highly specific restriction to forest trees and to New Zealand. We can generalise the query by replacing the specific terms by generic variables. But how do we generalise terms like "forest trees" and "New Zealand"? Let's suppose that we express the query using XML. Then the start of the query might look like this:

```
<query name="NZKAURI">
<tree>kauri</tree>
<country>New Zealand</country>
   … rest of definition …
</query>
```

The way to generalise the query is now obvious. We simply remove the restrictions and turn the whole thing into a function NZKAURI (topic, country). This function now has the potential to address questions on a much broader basis. In principle, it could answer questions about (say) kauri trees in other countries, other trees in New Zealand, and potentially about any kind of tree in any country.

Of course, the success of this kind of generalisation depends on exactly how the query is implemented. For instance, if it uses a number of isolated resources that refer only to New Zealand kauri trees, then it will fail completely for any other tree, or for any other country. The generalisation is most likely to be successful if the entire system confines itself to a fairly narrow domain and if it works with services that have a widespread coverage.

Generalised pattern matching, as described above, is just one of many ways in which an online agent can acquire information. Another possibility is provided by research into natural language processing. A system might take natural language queries entered by the user and map them into a specific search dialect.

11.2.5 The ant model of distributed intelligence

One of the strongest insights to flow from research in the new discipline of artificial life is that order can emerge in a system without any central planning or intelligence. In many living systems, order emerges instead through the interactions of many agents interacting with their environment and with each other. For instance, in many insect colonies, such as ants and bumblebees, the individual insects have no overall concept of what their colony should look like. Instead they behave according to very simple rules. For instance, if you are an ant and you see a scrap of waste lying around, then you pick it up. If you are carrying waste and find some waste lying around, then you drop the scrap that you are carrying. This simple action, repeated thousands of times, is all it takes to sort the contents of an ant colony into different areas for food, for eggs, waste, and so on.

Many useful applications flow from the above simple observation. For instance, Rodney Brooks applied the idea to robotics and created very effective mechanisms for controlling (say) robot walking, without any central control at all. Likewise the ant sort algorithm is a method by which computer systems can create order spontaneously, and without the need to applying specific sorting algorithms. In the ant sort, the ants either change or move items around, or if that is not possible, they can make it easier for other ants to find the same items again by leaving a trail of virtual pheromones.

So how do these ideas apply to online GIS? First, it is important to appreciate that the ant sort is an ideal method to apply on the Internet. The Web contains virtual oceans of data items, with more being added all the time. Suppose that virtual ants are looking for online items on a particular topic, say geographic data. Then they can "mark their trail" with virtual pheromones. These can take several forms. For example, one approach is to record a history of sites that you've visited with a score next to each one. Other ants looking at this list can see which sites have proved most relevant and useful.

There are difficulties with the ant sort model. One is that it applies best to a non-renewable resource. When real-life ants follow a pheromone trail, they take food away from the source that the trail leads to. When the source is exhausted, they stop laying down pheromones, so the trail goes cold. Not so an online resource. So there is a risk of concentrating huge amounts of activity on a single web site. One potential solution to this problem is to have the ant copy the information to a cache, and delete it if it is not accessed within a particular period.

11.3 MOBILE COMPUTING

Spatial information is an intricate part of our lives. From finding a vegetarian restaurant within walking distance of a hotel in a strange city to a garage which can repair a vintage Jaguar, mapping queries have so many, unexplored applications.

At present geographic information systems are the provinces of a minority, requiring special skills and software.

In the last year, cut-down GIS (Geographic Information Systems) have become available on palmtop computers, such as Tadpole, and FieldWorker. They operate on the model of providing GIS functionality on small data sets with upload/download options. Meanwhile mobile phones are starting to invade the Internet. But wireless devices have distinct limitations: they have small, limited colour screens and very low bandwidth. Thus an alternative is to think, not in terms of a micro-GIS, but in terms of online spatial queries.

A new approach is needed in which *spatial queries* are answered with information packed down by artificial intelligence on the server. Client and spatial profiling are essential for the best results, raising questions of privacy. The client, of course, does not wish to buy large digital maps (data sets) for his simple restaurant query, but wants to pay much less, say the cost of an information line (1900) phone call.

Handheld devices, such as mobile phones, have very limited facilities for human–computer interaction: keyboards are virtually non-existent; some sort of pen system might be available; screens are low resolution, usually monochrome. Voice recognition has great potential but would have to be inside the phone itself to allow really effective learning of an individual's voice. Thus queries must be brief and highly, but intuitively, codified. In addition the query needs include information that either identifies the user and device and evokes an associated profile, or provides a specification as in the emerging web standard for *Composite Capability/Preference Profiles* (CC/PP).

Privacy is a crucial issue. The more a data server knows about a client, the more it can tailor its response. However, the client is thereby giving up personal or organisational information, perhaps unwittingly. New privacy laws, such as those which came into force NSW in July 1, 2000, ensure the client has much greater access and control. A client needs assurance that his personal information will not be spread around the Internet if he entrusts it to some agent, for which mechanisms do not yet adequately exist. It is essential that the client has access and control of the data stored and there are mechanisms to prevent its unauthorised transfer. The data broker agent model enables personal information to reside within a trusted agent which can then formula queries in a sanitised fashion.

Large mapping organisations are under increasing pressure to sell their data in more ways. Online supply is obviously very promising, however, given the volume and diversity of data, an effective commercial model is needed which considers issues of tracking royalties through the flow of ownership, data watermarking, discounts for customer categories such as government and issues of resale of data by third parties.

New standards are emerging that will make all of the above possible. In addition to the OpenGIS specifications, which we described in Chapter 7, several specifications are under development by the Wireless Applications Forum. On the Web front there has been significant progress in digital signatures and security which we discussed briefly in Section 8.3.

11.4 FROM ONLINE GIS TO VIRTUAL WORLDS

Question: What's better than looking at a map of where you want to go? The answer is obvious: being there. But if you can't be there in person, then the next best thing is to be virtually there.

When an architect wants to show clients what a new development will look like, the building plans are only part of the story. Any major work includes artist's drawings of what the place would look like, and even models that clients can explore and look at from different angles.

In recent times, architects have begun making virtual reality models of the buildings they plan to build. That way, the owners can experience what the finished product will be like before building even begins. They can also try out design variations and potential colour schemes. The same kind of technology has been used to build virtual reconstructions of ancient buildings.

The idea of virtual space has been applied to GIS too, Dirk Spenneman, a colleague of ours, built an online virtual field trip around a GIS. The system allows students to get experience at planning fieldtrips before they undertake real fieldwork later on. One advantage is that they can learn the consequences of mistakes (e.g. failing to allow for poor weather) and avoid making them for real.

These kinds of technology can be extended to add virtual reality capability to GIS. Good examples of this approach are systems that have been developed to help people assess environmental impacts of logging and other activities. The program Smart Forest, for instance, combines a GIS, with forest models and virtual reality (Orland 1997). It allows users to define various scenarios and to experience the consequences.

The idea behind the program is to provide a sound basis for assessing the claims about environmental impact that are made by forestry companies and other groups. At present, a forestry company may have to make arguments such as "Well, we have to cut down part of the forest, but it will not spoil the aesthetics of the area and the patch will recover within twenty years." Instead of just saying this, the argument has much more force if people can actually see what the area will look like. This is exactly what Smartforest is intended to do (Orland 1997).

The program combines GIS, simple forest models, and virtual reality graphics. Once a model has been set up, the user can move around in the environment and look at the view from any place, in any direction, at any height from ground level to hundreds of metres.

Several other commercial packages of a similar nature are now available, such as Virtual Forest (Buckley et al. 1998).

Many computer games place the user in a virtual world. In games such as *SimCity* and *SimEarth*, for instance, the user acts as manager for entire cities or even the entire planet. The games are based around a GIS that allows the user to select regions and zoom in to the level of individual buildings. In other games, players can drive through virtual landscapes following road maps (e.g. *GranTourismo*) or negotiate their way through entire fictional 3D artificial worlds (e.g. *Tomb Raider*).

Having combined GIS and virtual reality, it is just a short step (conceptually at least!) to do the whole thing online.

Several projects have put this into practice (e.g. Fig. 11.2). Perhaps the best known is AlphaWorld (Active Worlds.com 2000). Participants can join a virtual community inhabited by "avatars", which are graphical representations of both themselves and other users. Not only can users move around and interact in this cyberworld, they can even build themselves virtual homes. Other online projects of this kind place the avatars in other environments, such as the surface of Mars. The Virtual Worlds Movement aims to develop online virtual reality as a means of developing virtual communities and other activities, such as business meetings.

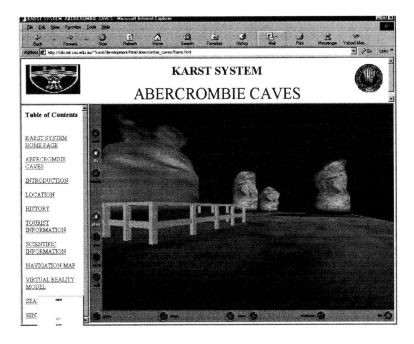

Figure 11.2. A virtual reality scene of the Abercrombie caves, New South Wales. The service, which is maintained by Charles Sturt University, can be accessed at http://clio.mit.csu.edu.au/

But if Internet users can explore alternate universes online, and interact with other explorers in the process, then why not virtual reproductions of the real thing? Many areas, such as city centres, are already mapped in great detail. By that we mean that the exact form of the buildings are well-known. Tourist maps of (say) downtown London sometimes include pictures of the buildings. Why not turn these picture maps into online virtual worlds. Tourists or students, for instance, could explore an area to get the feel of where things are before they even set foot there, or explore remote areas, such as the Andes or Himalayas, in greater depth than they could ever do in real life.

Glossary

Numbers following each definition indicate the chapter and section in which the relevant discussion is to be found.

Abstract specification Part of the OpenGIS model which is interface and implementation independent. 7.2

Access log A record kept by a web server of who down loads what. 3.2.3

Active Object Map A table maintained by the basic or portable object adaptors of objects on the bus or ORB. 7.3

ANZLIC Australia–New Zealand Land Information Council. 9.1

ANZMETA Top-level element in the ANZLIC metadata DTD. 9.1

Attributes Attributes appears in two different contexts: in XML or SGML it refers to properties of an element given values inside the start tag; in object-oriented design it refers to a property of an object, often a data value. 5.2.

Base map A map of a region on which various thematic layers are placed to create the final map. 1.2.1

Basic Object Adaptor (BOA) The first CORBA object adaptor which is responsible for activating objects on a server, superseded by the portable object adaptor (POA). 7.3.5

CEN Comité Européan de Normalisation. 9.3

CERCO Comité Européan des Responsables de la Cartographie Officielle. 9.3

CGI Common Gateway Interface, the link between a Web server and processing software. 3.3.1

Class A category of objects for data and processing. 1.2.3

Client-server protocol A communications model in which a *client* computer requests data from a *server* computer. 2.2.1

Client-side processing Actions performed on the computer where the Web browser is running. 4.1

Collection service (CORBA) The service for managing collections of objects. 7.3

Common Object Request Broker Architecture CORBA is a standard of the Object Management Group for activating and managing objects distributed over computer networks such as the Internet. 7.3.1

Component A modular software building block comprising a number of interrelated objects.

Concurrency The simultaneous running of more than one process. 7.3

Connectionless interaction Client server exchange in which the client does not form an on-going link to the server. 2.6.3.1

CSDGM Content Standard for Digital Geospatial Metadata. 9.2

Data archive A repository facility for storing data sets. 6.4.4

Data mining Techniques for knowledge discovery in huge datasets. Often associated with data warehouses. 10.7

Data warehouse An organised collection of databases and processes for information retrieval, interpretation and display. 6.3

DMG Data Mining Group. 10.9.3

DSTP Data Space Transfer Protocol. A standard for distributing data sets across the Internet. 10.9.4

DCMI The Dublin Core Metadata Initiative. A metadata initiative arising out of the Dublin Core workshop providing a series of metadata elements for authorship and other properties of documents. 8.2

Decision tree A computing model in which test conditions determine the branching and the leaves (ends of branches) contain definitions for values of target variables. 10.7

Digital Elevation Model (DEM) A model that specifies elevation for any point in a landscape. 1.2.1

Distributed data warehouse A data warehouse that is spread across several sites on the Internet. 6.3

Distributed objects Processing objects that interact across several sites on the Internet. 1.6

Document root (1) The base directory under which all documents reside on a Web server. 5.2

(1) The top of the document tree (see below) which in a DTD corresponds to the outermost (top level) element. 5.2

Document Tree The order and hierarchy of the elements of a document represented as a tree. 5.2

Document type definition (DTD) A meta-document of SGML that defines the elements and entities of any document instance and the structure to which they must conform. A DTD is analogous to a very general sort of template or a class in object oriented programming. 5.7.1

Dublin Core A standard for defining metadata within HTML documents (see also DCMI). 8.1

Dynamic access Access that is controlled at run-time. 7.3.2

Element A distinct component (e.g. section) in the structure of an SGML/XML document. Elements are the primary building block of SGML and XML documents. An element may contain other elements. 5.2

Enterprise Java Beans (EJB) The Java specification for large-scale distributed objects or frameworks. EJB has now moved to adopt the CORBA model, but is specific to the Java language. 7.3.6

Entity A property of an element. 5.2

Also, a syntactic variable that defines terms within a Document Type Definition. 5.2

Entities, public Entity sets external to a document, usually in some well-defined place, used for example for defining special characters. 5.2

ENV Euro-Norm Voluntaire. a European data standard. 9.3

Error log A record kept by a web server of unsuccessful attempts to access information. 3.2.3

Event service, CORBA A mechanism that allows components on the bus to register their interest in specific events. 7.3.7

Extensible Markup Language (XML) A system for marking up documents and data using tags that indicate structural elements. XML is a recommendation of the W3C for marking up documents, a later variant of SGML. 5.1

Externalisation service, CORBA A mechanism for streaming data in and out of components. 7.3.7

FGDC Federal Geospatial Data Committee (USA). 9.2.2

First order questions Questions about single objects or spatial variables (cf. second order), e.g. Where are parks located? 1.2.2

Framework A run-time system in which components are embedded.

Geographic Information System (GIS) A computer system for storing, displaying and interpreting geographic information. 1.1

Geographic Markup Language (GML) A system of XML tags for marking up geographic information. 10.9.2

GET A method of transferring form data to a server by embedding the data within a URL (web address). 3.3.1

GIE Group d'Interêt Economique. 9.3

GIS See Geographic Information System.

GHRC Global Hydrology Research Center. 10.8.1

GML See Geographic Markup Language.

Head An element at the start of an HTML document that contains metadata and other basic information about a document. 8.3

Hypertext Transfer Protocol (HTTP) The client-server protocol used in transferring data between servers and clients on the World Wide Web. 8.3

Hypermedia Information that combines multimedia with hypertext. 3.1.2

Hypertext Markup Language (HTML) A system of tags (based on an SGML DTD) used for marking up documents for the World Wide Web. 5.1

IDL Interface Definition Language. The fundamental building block of CORBA, which defines the interfaces to objects in a language independent way. 7.3.2

Imagemap An image within an HTML document within which different locations provide different hyperlinks. 4.2.1

Image input field An element within an HTML form that allows the user to input coordinates by clicking on an image. 3.3.2

Implementation repository A database on the ORB which contains information about object implementations. 7.3

INFO2000 European initiative on information sharing. 9.3

Information Network A set of sites on the Internet that collaborate to provide information about a particular theme or topic. 6.1

Inheritance Attributes and methods that a specialised class of objects inherit from a more general class of objects. 1.2.3

Instance A particular object that belongs to a given class. 1.2.3

IR Interface Repository. 7.3

Interface A specification of the procedure and properties of an object abstracted away from how the object is implemented. 7.3

Interface Repository (IR) A database on the ORB which contains information about object interfaces. 7.3.2

Internet A global computer network, linked via the Internet Protocol (IP) and governed by the Internet Society. 1.3

Java A programming language developed by SUN Corporation to run processes on a Web client machine. 7.3.6

Javascript A scripting language developed by Netscape to allow processing within HTML documents. 4.4

JPL Jet Propulsion Laboratory. 10.8.1

Kriging A statistical technique used to interpolate values of spatial variables between points where observations are available. 1.2.2

Layer A set of geographic information, usually dealing with a single theme, that can be combined with other layers to form a map. 1.2.1

Legacy In the CORBA context, refers to pre-existing software and data. 7.3

Legacy data Datasets from old or historic sources. 7.3.1; 10.3.1.

Licensing service (CORBA) The service which allows monitoring of an object's activities for accounting purposes. 7.3.7

Markup A procedure for indication document structure or format through the use of textual tags embedded within the document. 5.1

Mark up tags Labels inserted into a document or data set to indicate structure, formatting or processing. 5.1

MEGRIN Multipurpose European-Ground Related Information Network. 9.3

Metadata Literally, data about data. Information about the provenance and nature of information resources. 6.4.1; 6.5.2.1

MetaLite A simplified version of the FGDC metadata standard used in parts of the United States. 9.2.5

Methods The computational procedures (subroutines) used to access and modify the properties (data) of an object. 1.2.3; 7.3.1

Mirror site A complete copy of an information resources that is held at a separate Web site. 6.4.4

Namespace A collection of terms and definitions that describe the usage and sometime meaning of XML tags (i.e. the semantics of the markup language). 5.4

Naming service (CORBA) A means for providing names for objects. 7.3.7

NGDCH National Geospatial Data Clearing House (USA). 9.2.2

NCDM National Center for Data Mining. 10.8.3

NCSA National Center for Supercomputer Applications. Their development of the first multimedia Web browser sparked rapid growth of the World Wide Web in the early 1990s. 1.3

NOAA-SAA National Oceanic and Atmospheric Administration-Satellite Active Archive. 10.8.1

NSIDC National Snow and Ice Data Center. 10.8.1

NSDI National Spatial Data Infrastructure. 9.2.2

Nuggets Items of information uncovered by data mining. 10.7

Object Management Group (OMG) A large software and hardware vendor consortium aiming to provide industry wide standards for object technology. 7.1

Open Software Foundation (OSF) A vendor consortium developing operating system and process communication standards. 7.3.2

ORNL Oak Ridge National Laboratory. 10.8.1

OMG Object Management Group. 7.1; 7.3.1

OQL Object Query Language. 5.6

Object-oriented An approach to information processing based on objects and their relationships. 1.2.3

OCLC Online Computer Library Center. 8.2

Object Query Language (OQL) A proposal for extracting data from objects in a collection of some kind. 5.6

Object Request Broker (ORB) A fundamental building block of CORBA which allows objects to interact across a network. 7.3

OSF Open Software Foundation. 7.3.2.2

Overlay The act of laying one thematic layer over another layer or base map. 1.2.2

Path The sequence of directories and processes that need to be invoked to reach an item of information. 3.3.1

PERL A freeware scripting language that is widely used to implement processing associated with Web services. 3.2; 3.5

Persistence service (CORBA) A means of making objects appear as if they are always available online (when in fact they may be transparently archived to databases or other file systems). 7.3.7

Pixel A point within an array that combines to form an image. 1.2.1

Portable Object Adaptor (POA) The successor to the Basic Object Adaptor. 7.3.4; 7.3.5

Post method Method of transmitting data from Web forms in which the data are encapsulated and sent via HTTP as standard input to a Web server. 3.3.1

PMML Predictive Model Markup Language. An XML based language that is used to describe models. 10.9.3

Pragma A compiler directive. 7.3.2

Quadtree An indexing method that progressively divides a landscape into quarters, subquarters and so on. 1.2.1

Quality The completeness, correctness and relevance of data. 1.3; 6.5.2

Query service Queries for objects based on SQL and the OMG's Object Query Language. 7.3.7

Raster layers Map layers that contain arrays of pixel data (e.g. satellite images). 1.2.1

RDF type Indicates an RDF composite construct such as a collection. 8.5.3

Reference concrete syntax Part of the SGML specification which specifies the characters and restrictions of markup text, e.g. the use of <> in defining beginning and end tags. 5.2

Relationship A link between objects. There are three kinds. (a) Common attributes or methods; (b) GenSpec, in which one object is a special case of a more general one. (c) Whole-Part, in which one object forms part of another. 1.2.3.

Repository Id A unique identifier for an object in a repository, valid across repositories and networks. 7.3.2

Request for Proposal (RFP) (1) An official Internet Society announcement seeking proposals relating to a particular standard or protocol. ALSO

(2) A tendering process used by organisations such as the Open GIS Consortium as the first stage in developing a standard. 7.3.7

RDF The Resource Description Framework. A W3C standard for metadata describing online resources, such as sites and services. 8.1

Resource Discovery The process of searching for online information and services 8.2

Scalable Vector Graphics (SVG) An XML based language for defining vector images. This standard provides images that can be rescaled without distortion. 4.4.5

Schema An XML document describing the structure and semantics of the class of XML documents which use it. 8.1

Second order questions Questions about relationships between geographic variables (e.g. what area of forest lies within parks). 1.2.2

Security service A range of operations for controlling access, authentication etc. 7.3.7

Semantics The meaning of terms and expressions in a language. 5.1; 8.2

Serialisation syntax The representation of an RDF model in XML. 8.5.4

Servants The user written objects on a server which interact with the server skeletons. 7.3.5

Server A program that transmits data in response to requests from clients. 4.3.4

Server-per-method A server activation protocol in which a server process is started for each activation of an object method. 7.3.5

Shared server A server activation protocol in which many object methods share a single (multi-threaded) process. 7.3.5

Skeleton An automatically generated code fragment which interfaces the user's server-side code (servant) with the ORB. 7.3.2

Spatial Reference System A coordinate system for points on the earth, defining units, transformations and reference points. 7.3.9

Stability The capacity of online sites and services to keep functioning, and to remain at the same Web address. 1.3; 6.4.4

Standard Generalised Markup Language (SGML) A general approach to marking up the structure and format of text according to a document type definition (DTD). HTML is one example. XML is a simplified variant of SGML. 5.1

Standardisation The process of making information and services conform to standards. 1.3

Standards Precise specifications for particular operations or structures. 6.4.1

SVG See Scalable Vector Graphics.

SWG Standards Working Group. 9.2.2

Stateless Any process in which no state (i.e. a record of history) is recorded. 2.6.3

Theme A set of geographic information with a common topic, or related features. 1.1

Thematic layer A map layer with a particular theme (e.g. roads). 1.1

Time service A service that provides time information and synchronisation in a distributed system. 7.3.7

Trading service The "yellow pages" of a distributed object system, allowing an object to be located on a network on the basis of its properties and function. 7.3.7

Transaction service The means of controlling concurrent access to data files. 7.3.7

Unshared server A server policy in which a server is created for each and every object. 7.3.5

URL Uniform Resource Locator. 3.3.1

VBScript Visual Basic Script, a language for implementing data processing operations on personal computers. 4.3.6

Vector layer A map component consisting of points, lines and polygons. 1.2.1

Virtual library An information service consisting of hypertext links to online information on a range of topics. 6.3

W3C The World Wide Web Consortium, the body which manages standards for the Web. 1.3

Web Client A computer program that retrieves information from World Wide Web servers. 3.1

Web server A server that delivers information in response to HTTP requests across the Internet. 3.1

Whole-part relationship A situation in which one object forms a part of another object (e.g. town objects form part of a landscape object). 1.2.3

XML See Extensible Markup Language. 5.1

Bibliography

ActiveWorlds.com (2000). *Alpha World.* http://www.activeworlds.com/

Aha, D. (1995). *Machine Learning.* Tutorial on Machine Learning. AI and Statistics Workshop. Ft Lauderdale, Florida.

Alschuler L. (1995). *ABCD...SGML.* International Thomson Publishing, Boston

ANZLIC (2000). *Australasian Spatial Data Directory.* http://www.environment. gov.au/net/asdd/

ANZLIC Home Page. http://www.anzlic.org.au/

Booch, G., Rumbaugh, J. and Jacobson, I., (1999). *The (U)nified (M)odelling (L)anguage User Guide.* Addison–Wesley, Reading, Massachusetts.

Bossomaier, T.R.J. and Green, D.G. (2000). *Complex Systems.* Cambridge University Press, Cambridge.

Bossomaier, T.R.J. and Green, D.G. (2001). *Spatial Metadata and Online GIS Website.* http://www.csu.csu.edu.au/complexsystems/smdogis/

Brunsdon, C., Fotheringham, A.S. and Charlton, M.E. (1996). Geographically weighted regression: A method for exploring spatial non-stationarity. *Geographical Analysis* 28, 281–298.

Bryan, M. (1988). *SGML: An Author's Guide to the Standard Generalized Markup Language.* Addison-Wesley, Reading, Massachusetts.

Buckley, D.J., Ulbricht, C. and Berry, J. (1998). *The Virtual Forest: Advanced 3-D Visualization Techniques for Forest Management and Research.* ESRI User Conference, July 27–31, 1998 San Diego, CA. http:// www.innovativegis.com/ products/vforest/

Burdet, H.M. (1992). What is IOPI? *Taxon* 41, 390– 392. http://life.csu.edu.au/iopi/

Buttenfield, B.P. (1998). Looking Forward: Geographic Information Services and Libraries in the Future. *Cartography and GIS* 25(3), 161–171.

Cathro, W. (1997). *Metadata: An Overview.* Standards Australia Seminar, August 1997. http://www.nla.gov.au/nla/staffpaper/cathro3.html

Chawla, S., Shekhar, S., Wu, W.L. and Ozesmi, U. (2001). Modeling spatial dependencies for mining geospatial data: An introduction. In H.J. Miller and J. Han (eds) *Geographic Data Mining and Knowledge Discovery.* London, Taylor and Francis (in press).

Clark, J. (1999). *XML Namespaces.* http://www.jclark.com/xml/xmlns.htm

Cliff, A.D. and Haggett, P. (1998). On complex geographical space: Computing frameworks for spatial diffusion processes. In P.A. Longley, S.M. Brooks, R. McDonnell and B. MacMillan (eds) *Geocomputation: A Primer* (Chichester, U.K., John Wiley and Sons), pp. 231 –256.

Danko, D.M. (1992). The Digital Chart of the World Project. *Photogrammetric Engineering & Remote Sensing* 58(8), 1125–1128.

DMA (Defense Mapping Agency) (1992). *Digital Chart of the World.* Defense Mapping Agency, Fairfax, Virginia. (Set of four CD-ROMs.) http://edc.usgs.gov/glis/hyper/oldguides/dcw

Dietterich, T.G. (1996). Machine Learning. *ACM Computing Surveys.* 28(4es), December.

DMG (2000). PMML 1.0 – Predictive Model Markup Language. Data Mining Group (DMG). http://www.dmg.org/html/pmml_v1_1.html

Drew, P. and Ying, J. (1998). Metadata management for geographic information discovery and exchange. In Sheth, A. and Klas, W. (eds), *Multimedia Data Management: Using Metadata to Integrate and Apply Digital Media* (McGraw–Hill), pp. 89–121.

Dublin Core. *Dublin Core Metadata Initiative* (homepage). http://purl.org./dc/

Eklund, P. W., Kirkby, S. D. and Salim, A. (1998). Data mining and soil salinity analysis. *International Journal of Geographical Information Science* 12, 247–268.

Ensign, C. (1997). *SGML: The Billion Dollar Secret.* Prentice Hall, New Jersey.

Erdos, P. and Renyi, A. (1960). On the Evolution of Random Graphs, *Math. Inst Hungarian Acad*, 5, 17–61 (in Hungarian).

ERIN (1995). *Species Mapper.* http://www.environment.gov.au/ (now offline).

ERIN (1999). *Environment Australia* http://www.environment.gov.au/

Ester, M., Kriegel, H-P. and Sander, J. (1999). Knowledge Discovery in Spatial Databases Invited Paper at *23rd German Conference on Artificial Intelligence* (KI '99). Bonn, Germany, 1999.

Etzioni, O. (1996). The World Wide Web – Quagmire or Gold Mine? *The Communications of the ACM.* November (39)11, 65–68.

European Petroleum Survey Group (2000). *Petrochemical Open Software Consortium.* http://www.epsg.org/

Evans, C., Feather, C.D.W., Presler-Marshall, M., and Resnick, P. (1997). *PICSRules 1.1.* W3C. http://www.w3.org/TR/REC-PICSRules

Fayyad, U.M., Piatetsky-Shapiro, G., and Smyth, P. (1996a). The KDD Process for Extracting Useful knowledge from Volumes of Data. *The Communications of the ACM.* November 39(11), 27–31

Fayyad U.M., Piatetsky-Shapiro G., and Smyth P. (1996b). From Data Mining to Knowledge Discovery: An Overview. In: *Advances in Knowledge Discovery and Data Mining.* AAAI Press, Menlo Park, 1996, pp. 1–34.

Ferraiolo, J. (2000). *Scalable Vector Graphics (SVG) 1.0 Specification.* W3C Candidate Recommendation. http://www.w3.org/TR/2000/CR-SVG-20001102/

Gamma, E., Helm, R., Johnson, R. and Vlissides, J. (1995). *Design Patterns: Elements of Reusable Object-Oriented Software.* Addison-Wesley, Reading Massachusetts.

Gardner, C. (1996). *IBM Data Mining Technology.* Stamford, IBM Corporation, Connecticut.

Garfinkel, S., (1995). *PGP: Pretty Good Privacy.* O'Reilly & Associates, Sebastopol, CA.

Giarrantano, J. and Riley, G. (1989). *Expert Systems: Principles and Programming*, Boston: PWS-KENT Publishing. http://www.ghgcorp.com/clips/CLIPS.html

Goldfarb, C.F. and Prescod, P. (1998). *The XML Handbook.* Prentice Hall, N.J.

Green, D.G. (1993a). *The Guide to Australia.* Charles Sturt University. http://www.csu.edu.au/australia/

Green, D.G. (1993b). Emergent behaviour in biological systems, In Green, D.G. and Bossomaier, T.R.J. (eds), *Complex Systems – from Biology to Computation.* pp. 24–35, IOS Press, Amsterdam

Green, D.G. (1994). Databasing diversity – a distributed public-domain approach. *Taxon* 43, 51–62.

Green, D.G. (1995). From honey pots to a web of SIN – building the world-wide information system. In Tsang, P., Weckert, J., Harris, J. and Tse, S. (eds), *Proceedings of AUUG'95 and Asia-Pacific World Wide Web '95 Conference*, Charles Sturt University, Wagga Wagga, pp. 11–18. http://www.csu.edu.au/special/conference/apwww95/papers95/dgreen/dgreen.html

Green, D.G. and Croft, J.R. (1994). Proposal for Implementing a Biodiversity Information Network. In Canhos, D.A.L., Canhos, V. and Kirsop, B. (eds), *Linking Mechanisms for Biodiversity Information.* Proceedings of the Workshop for the Biodiversity Information Network, pp. 5–17. Fundacao Tropical de Pesquisas e Tecnologia "Andre Tosello", Campinas, Sao Paulo, Brazil.

Green, D.G. (1996). A general model for on-line publishing. In: Bossomaier, T. and Chubb, L. (eds), *Proceedings of AUUG'96 and Asia-Pacific World Wide Web '96 Conference*. Australian Unix Users Group, Sydney. pp. 152– 158.

Green, D.G. and Klomp, N. (1997). Networking Australian biological research . *Australian Biologist* 10(2), 117– 120.

Green, D.G. Bristow, P., Ash, J., Benton, L., Milliken, P., and Newth, D. (1998). Network Publishing Languages. In Helen Ashman & Paul Thistlethwaite (eds), *Proceedings of the Seventh International World Wide Web Conference* . Elsevier, Amsterdam. http://life.csu.edu.au/~dgreen/papers/www7.html

Green, D.G. (2000). Coping with complexity – the role of distributed information in environmental and resource management. In Salminen, H., Saarikko, J., and Virtanen, E. (eds), *Resource Technology '98 Nordic – Proceedings*. Finish Forestry Research Institute, Rovaniemi, Finland.

Hammer, J., Garcia-Molia, H., Labio, W., Widom, J. and Zhuge, Y. (1995). The Stanford Data Warehousing Project. *Data Engineering Bulleting. Special Issue on Materialized Views and Data Warehousing* 18(2), 41– 48.

Hardy, G. (1998). *The OECD's Megascience Forum Biodiversity Informatics Group*. http://www.oecd.org//ehs/icgb/BIODIV8.HTM

Hawkins, H.S., Rimmington, G.M. and Peter, I. (1992). LandcareNET – A new medium for agricultural communication. *Agricultural Science* 5(2), 35– 40.

Hubber, H. (1997). A Success Story: Wal-Mart Stores. In *Proceedings of the First State of Florida Data Warehousing Conference.* (UNPUBLISHED)

Inmon, W.H. (1995). What is a Data Warehouse? *Prism* (1)1.

Inmon, W.H. (1996). The Data Warehouse and Data Mining. *Communications of the ACM*. November 39(11), 49–50.

ISOC (2000). The Internet Society (ISOC), Home Page. http:// www.isoc.org/

IUBS (1998). Species 2000. International Union of Biological Sciences. http://www.sp2000.org/

IUFRO (1998). *Global Forest Information Service* . International Union of Forestry Research Organisations. http://iufro.boku.ac.at/

Jacobson, I., Griss, M. and Jonsson, P. (1997). *Software Reuse*. ACM Press, New York.

Kennedy, R.L., Lee, Y., van Ray, B., Reed, C.D., and Lippman, R., (1998). *Solving Data Mining Problems Through Pattern Recognition* . The Data Mining Institute and Prentice Hall.

Knuth, D.E., (1984). *The TeX Book* . Addison-Wesley, Massachusetts.

Koperski, K., Han, J. and Adhikary, J. (1999). Mining knowledge in geographic data. *Comms. ACM.* http://db.cs.sfu.ca/sections/publication/kdd/kdd.html

Krauskopf, T., Miller, J., Resnick, P. and Treese, W. (1996). *PICS Label Distribution, Label Syntax and Communication Protocols: Version 1.1* , W3C, http://www.w3.org/TR/REC-PICS-labels

Krol, E. (1992). *The Whole Internet User Guide and Catalog*. O'Reilly & Associates, Sebastopol CA.

Lamport, L. (1986). *LaTeX: A Document Preparation System.* Addison-Wesley, Massachusetts.

Larman, C. (1998). *Applying UML and Patterns*. Prentice Hall, New York.

Lassila, O. and Swick, R.R. (1998). *Resource Description Framework (RDF) Model and Syntax*. World Wide Web Consortium. (Online) http://www.w3.org/TR/1998/WD-rdf-syntax-19980216.

Lassila, O. and Swick, R.R. (1999). *Resource Description Framework (RDF) Model and Syntax Specification.* W3C. http://www.w3.org/TR/REC-rdf-syntax.

Lees, B.G. and Ritman, K. (1991). Decision-tree and rule-induction approach to integration of remotely sensed and GIS data in mapping vegetation in disturbed or hilly environments. *Environmental Management* 15, 823–831.

Lonely Planet Publications (2000). *Lonely Planet Travel Guides.* http://www.lonelyplanet.com/

Lu, H., Setiono, R., and Liu, H. (1996). Effective data mining using neural networks. *IEEE Transactions on Knowledge and Data Engineering* 8(6), 957 –961.

Malhotra, A. and Maloney, M. (1999). *XML Schema Requirements*. W3C http://www.w3.org/TR/NOTE-xml-schema-req

MacEachren, A.M., Wachowicz, M., Edsall, R., Haug, D. and Masters, R. (1999). Constructing knowledge from multivariate spatio-temporal data: integrating geographical visualization with knowledge discovery in database methods. *Intern. J. Geogr. Information Science* 13(4), 311–334.

Malerba, D., Esposito, F. Lanza, A., and Lisi., F.A., (2001). Machine learning for information extraction from topographic maps. In H.J. Miller and J. Han (eds), *Geographic Data Mining and Knowledge Discovery*, (London, Taylor and Francis) (in press).

Mesrobian, E., Muntz, R., Shek, E., Nittel, S., La Rouche, M., Kriguer, M., Mechoso, C., Farrara, J., Stolorz, P. and Nakamura, H. (1996). Mining geophysical data for knowledge. *IEEE Expert* 11(5), 34–44.

Michalski, R.S., Bratko, I., and Kubat, M. (1998). *Machine Learning and Data Mining Methods and Applications.* John Wiley. New York.

Miller, J., Resnick, P. and Singer, D. (1996). *Rating Services and Rating Systems (and their Machine Readable Descriptions): Version 1.1*, W3C, http://www.w3.org/TR/REC-PICS-services

Miller, H. and Han, J., (eds) (2001). *Geographic Data Mining and Knowledge Discovery*. Taylor and Francis, London.

NCDM (2000). *Data Space Transfer Protocol (DSTP)*. National Center for Data Mining, University of Illinois at Chicago (UIC). http://www.ncdm.uic.edu/dstp/

National Center for Supercomputer Applications (NCSA) (1995). *NCSA Imagemap Tutorial*. http://hoohoo.ncsa.uiuc.edu/docs/tutorials/imagemapping.html

NGDC (2000). WebMapper Interface. National Geophysical Data Center. http://www.ngdc.noaa.gov/paleo/

NISO (1999). The ANSI/NISO Z39.50 Protocol: Information Retrieval in the Information Infrastructure. National Information Standards Organisation.

OMG (1997). The Object Management Group (OMG) Home Page. http://www.omg.org/

Online Computer Library Center (OCLC) (1997). *Center Home Page*. http://www.oclc.org/

OGC (2000a). *OpenGIS® Abstract Specification*. Open GIS Consortium http://www.opengis.org/

OGC (2000b). *Geography Markup Language (GML) v1.0*. Open GIS Consortium http://www.opengis.org/

OSF (2000). *Open Source Foundation*, Home Page. http://www.opensource.org/

Openshaw, S., Cross, A. and Charlton, M., (1990). Building a Prototype Geographical Correlates Machine. *Intern. J. Geographical Information Systems*, 4(4), 297–312.

Orffali, R., Harkey, D. and Edwards, J. (1997). *Client/Server Survival Guide*, (3rd ed). John Wiley & Sons, New York.

Orland, B. (1997). Forest visual modeling for planners and managers. *Proceedings, ASPRS/ACSM/ RT'97, Seattle*. American Society for Photogrammetry and Remote Sensing, Washington vol 4, pp.193–203. http://www.imlab.uiuc.edu/smartforest/

Plewe, B. (1997). *GIS Online*. Onward Press, Albany New York.

Quinlan R.J. (1993). *C4.5 Programming for Machine Learning*. Morgan Kaufmann, New York.

Roddick, J.F. and Spiliopoulou, M. (1999). A bibliography of temporal, spatial and spatio-temporal data mining research. *SIGKDD Explorations* 1(1), 34–38. http://www.cis.unisa.edu.au/~cisjfr/STDMPapers/.

Schwartz, R.L. (1993). *Learning Perl.* O'Reilly & Associates, Sebastopol CA.

Steinke, A., Green, D.G. and Peters, D. (1996). On-line environmental and geographic information systems. In Saarenma, H. and Kempf, A. (eds), *Internet Applications and Electronic Information Resources in Forestry and Environmental Sciences.* EFI Proceedings No. 10. European Forestry Institute, Joensuu (Finland), pp. 89–98.

Srinivasan, A. and Richards, J.A. (1993). Analysis of GIS spatial data using knowledge-based methods. *International Journal of Geographical Information Systems* 7, 479–500.

Travis, B.E. (1997). *OmniMark at Work.* SGML University Press, Denver, CO.

United Nations Environment Programme (UNEP) (1995). Background Documents on the Clearing-House Mechanism (CHM). *Convention on Biological Diversity.* Jakarta Indonesia. http://www.biodiv.org/chm/info/official.html

US Census Bureau (1994). *The TIGER mapping system* . http://tiger.census.gov/

Wall, L. and Schwartz, R.L. (1991). *Programming Perl.* O'Reilly & Associates, Sebastopol CA.

Weibel, S., Kunze, J. and Lagoze, C. (1998). *Dublin Core Metadata for Simple Resource Discovery.* Dublin Core Workshop Series. (Online) http://purl.oclc.org/ metdata/dublin_core_elements/draft-kunze-dc-02.txt

Wessel, P. and Smith, W. H. F. (1991). Free software helps map and display data. *Eos Trans., American Geophysical Union* 72, 441.

Wessel, P. and Smith, W.H.F. (1995). New Version of the Generic Mapping Tools Released, *Eos Trans. American Geophysical Union* . http://www.agu.org/eos_elec/ 95154e.html

Whalen, D. (1999). *The Cookie FAQ.* Cookie Central. http://www.cookiecentral. com/faq/

Wiener J.L. (1997). *Data Warehousing: What is it? And Related Stanford DB Research.* Stanford Database Research Laboratory.

Worbel, S., Wettschereck, D., Sommer, E., and Emde, W. (1997). Extensibility in data mining systems. In Simoudis, E. and Han. J. (eds), *The Proceedings of The 2nd International Conference On Knowledge Discovery and Data Mining* . AAAI.

World Wide Web Consortium (W3C) (1999). *The Document Object Model.* World Wide Web Consortium. http://www.w3c.org/rdf

Xerox PARC (1993). *Map Viewer.* http://pubweb.parc.xerox.com/map

Zhuge, Y., Garcia-Molina, H., Hammer, J., and Widom, J. (1995). View Maintenance in a Warehousing Environment In *Proceedings of the ACM SIGMOD*

International   Conference   on   Management   of   Data   pp.   316–327.   San   Jose, California.

Index

Entries in bold indicate extended passages with multiple references over two or more pages. In many cases they comprise sections dealing with that topic.

Learning Resources
Centre